Chris Willmott

AF832453

Biological Determinism, Free Will and Moral Responsibility

Insights from Genetics and Neuroscience

Springer

Chris Willmott
Department of Molecular and Cell Biology
University of Leicester
Leicester
UK

ISSN 2211-8101 ISSN 2211-811X (electronic)
SpringerBriefs in Ethics
ISBN 978-3-319-30389-5 ISBN 978-3-319-30391-8 (eBook)
DOI 10.1007/978-3-319-30391-8

Library of Congress Control Number: 2016933111

© The Author(s) 2016
This work is subject to copyright. All rights are reserved by the Publisher, whether the whole or part of the material is concerned, specifically the rights of translation, reprinting, reuse of illustrations, recitation, broadcasting, reproduction on microfilms or in any other physical way, and transmission or information storage and retrieval, electronic adaptation, computer software, or by similar or dissimilar methodology now known or hereafter developed.
The use of general descriptive names, registered names, trademarks, service marks, etc. in this publication does not imply, even in the absence of a specific statement, that such names are exempt from the relevant protective laws and regulations and therefore free for general use.
The publisher, the authors and the editors are safe to assume that the advice and information in this book are believed to be true and accurate at the date of publication. Neither the publisher nor the authors or the editors give a warranty, express or implied, with respect to the material contained herein or for any errors or omissions that may have been made.

Printed on acid-free paper

This Springer imprint is published by Springer Nature
The registered company is Springer International Publishing AG Switzerland

For my parents, Peter and Ann,
in the year of their golden wedding

SpringerBriefs in Ethics

More information about this series at http://www.springer.com/series/10184

Preface

It is a sad reality that we live in a world in which people commit crime. Every day, in every city, in every country, the actions of one individual will impact detrimentally on the person or the property of somebody else.

As part of the commitment to living in civilised society, those accused of carrying out antisocial behaviour of sufficient gravitas will have the evidence against them considered in a court of law. If found guilty, they will be required to fulfil some suitable punishment.

Suppose, for a moment, that it could be shown that the person who had perpetrated a crime had been literally unable to avoid carrying it out. Under such circumstances, would it still be appropriate to punish them for their actions? As details regarding the influence of both genetic characteristics and brain neurochemistry on human behaviour are being uncovered, some scientists and philosophers are claiming that traditional notions of free will, of moral responsibility and, therefore, of accountability for one's actions need to be re-evaluated.

In order to investigate the legitimacy of these propositions, it is necessary to undertake a survey of various relevant discourses. First, we need to reflect upon philosophical considerations of free will and determinacy. Are, for example, free will and determinism mutually exclusive, or might there be some way in which biological determinism and moral responsibility might co-exist? This will be considered in Chap. 1.

Second, it is important to have an appreciation of the current legislation regarding responsibility for one's actions. This will be the focus of Chap. 2. Building on these philosophical and legal foundations, it will then be time in Chap. 3 to investigate the scientific discoveries which are leading some commentators to question the existence of free will and/or moral responsibility. Evidence drawn from both genetic analysis and brain science will be considered.

The past decade has seen an explosion of interest in the potential relevance of such brain-related science in legal cases, and the emergence of a new field of "neurolaw". A survey of examples in which genetic and neuroscientific data have already been used in criminal trials (Chap. 4) will precede a final discussion

(Chap. 5) in which these disparate threads will be interwoven into reflections on the validity of biological determinism as an influence in human behaviour, and the appropriateness of such genetic and brain imaging evidence in current and future criminal proceedings.

Acknowledgements

Thanks to Trevor Stammers and Matthew James for reading an earlier version of this book. I am grateful to Sally Cunningham for helpful criticism of Chap. 2 on legislation associated with mental disorders, and to Jon Scott and Blair Grubb for checking Sect. 3.1 on brain architecture. All errors, of course, remain my own. Thanks also to my family, particularly to Anne, for her faithful support.

Contents

Summary

Recent investigations have shed light on the roles played by genetics and neuroscience in human behaviour. These observations have led some commentators to adopt a model of biological determinism in which the role of free will is downplayed or entirely excluded.

Following an overview of philosophical aspects of free will and determinism, and a survey of current legislation relating to responsibility, the evidence for biological contributions to behaviour is reviewed. Whilst the importance of genetics and neurobiology to behaviour is generally endorsed, it is concluded that this connection does not currently substantiate a deterministic view in which moral responsibility for one's actions is undermined.

Current and potential future roles for biological evidence in criminal proceedings are considered. Despite the fact that genetic or brain imaging data have already been cited in some court cases, it is argued that expansion of this usage at the present time would be premature. As further data is gathered, however, a point may be reached at which biological information would have a role to play as mitigation during the sentencing phase of trials.

Chapter 1
Free Will and Determinism

Abstract In this opening chapter, we begin by outlining ways in which philosophers have sought to resolve the "free will problem". That is to say, how can the apparent existence of individual free will be reconciled with evidence that suggests the universe (or elements within it) are predetermined? A variety of different models are introduced, including Libertarianism (which holds that life is not determined and we do have free will), Hard determinism (in which life is determined and free will is illusory) and Compatibilism (in which room must be made for the cohabitation of the existence of both free will and determinism). In the context of the rest of the book it is hard determinism, and particularly biological determinism, which presents the greatest challenge to the attribution of moral responsibility and legal recompense. If aspects of my genetics or my brain has determined a course of action, and I was not therefore at liberty to do otherwise, can I be held culpable for that behaviour? This is an issue we will unpack in later chapters.

Keywords Alternative possibilities · Compatibilism · Determinism · Folk intuitions · Free will · Incompatibilism · Libertarianism · Moral responsibility · Self-forming actions · Ultimate responsibility

1.1 Folk Beliefs About Free Will

Over recent years, academics in the emerging discipline of "experimental philosophy"[1] have used a variety of approaches to investigate "folk intuitions" about free will, that is to say the innate beliefs ordinary people would hold on the subject (Sommers 2010). These studies have found strong support for the notion that human decision making is an integral component of free will.

For example, when given the opportunity to articulate their understanding of the concept of free will in their own words, 65 % of undergraduates surveyed made

[1]Experimental philosophy is characterised by conducting fieldwork to endorse philosophical claims, in contrast to more traditional philosophy which is termed "armchair philosophy" by practitioners of the more empirical persuasion.

© The Author(s) 2016

C. Willmott, *Biological Determinism, Free Will and Moral Responsibility*,
SpringerBriefs in Ethics, DOI 10.1007/978-3-319-30391-8_1

reference to the ability to make a decision or choice (Monroe and Malle 2010). When the same cohort was challenged with the suggestion that neuroscientists believed free will to be an illusion, 49 % refuted this proposal and a further 25 % prevaricated.

In a separate, questionnaire-based, study participants were presented with two theoretical universes; a deterministic one, in which "*everything that happens is completely caused by whatever happened before it*" and an indeterminist universe, in which "almost *everything that happens is completely caused by whatever happened before it. The one exception is human decision making*." (Nichols and Knobe 2007: both quotes from p. 669). Over 90 % of the respondents felt that the indeterminist universe was more akin to our own.[2]

This should not come as a surprise. Our capacity for free will seems intuitive. When I got up for breakfast this morning it seems that I was entirely at liberty to choose Corn Flakes instead of muesli (or indeed to have skipped straight to toast) and it would appear nonsensical to suggest that some other forces in the universe had influenced me to the extent that they had effectively made that selection on my behalf. How could they? Why would they? Similarly, was it not my free choice to spend several months researching and writing this book rather than filling my time in different ways?

1.2 Theological Objections to Free Will

Historically, the principal objections to the notion of human free will were theological in nature. If, it has been argued, God is all powerful (omnipotent) and all knowing (omniscient) can we be said to be truly free? Dartmouth College philosopher Adina Roskies has neatly summarised the predicament: "*God's omnipotence means that he could control our actions if he so wished, but that human freedom is preserved because he refrains from controlling us. However, God's omniscience presents a further problem for human freedom, for even if God does not control our actions, if he knows now how we will act before we act, then we are not free to do otherwise—foreknowledge seems to foreclose the possibility of freedom of the will, for our actions are predetermined.*" (Roskies 2006: p. 419).

Further discussion of theological consideration of predestination versus free will is beyond the scope of this present book, though interested readers are referred to *Divine Foreknowledge: Four Views* (Beilby and Eddy 2002).

[2]The study is actually much richer and more nuanced than this bold statement implies. Nevertheless, none of their findings contradicts the observation noted here, which is sufficient to support the key point being addressed in this introductory statement.

1.3 Philosophical Considerations of Free Will and Determinism

At first glance, notions of individual free will and a deterministic view of the universe would seem to be irreconcilable; if my actions have been entirely decided by external factors, in what way can I be said to have chosen to do something worthy of blame (or, indeed, praise)? However, philosophers throughout the centuries, going back at least as far as Chrysippus in the third century BCE, have expended substantial effort demonstrating that (some form of) freedom or autonomy and (some form of) determinism can coexist (Bobzien 2001).[3] Various arguments in support of this position will be considered below.

If free will and determinism are not direct opposites, then "*Does free will exist?*" and "*Is the universe deterministic?*" are therefore separate and distinct questions. In consequence, it becomes possible to draw up a matrix, such as the one given in Table 1.1, to map out the principal positions held by different philosophers.

The view that "free will" and a deterministic universe can be brought together is known as "compatibilism" (or "soft determinism", in order to distinguish it from "hard determinism", discussed below). Compatibilism (see Sect. 1.3.1) is a view held widely both by individuals who have given the question of free will a lot of thought, and by those acting more instinctively to try and reconcile different aspects of their experience of being human.

If someone is not persuaded that free will and a deterministic universe are compatible then it follows that they are an "incompatibilist". As is evident from Table 1.1, however, an incompatibilist position might be held for one of two divergent reasons. Firstly, if they contend that the strength of the case for free will rules out a deterministic universe (and/or the case for indeterminism is considered so overwhelming) then the person is a Libertarian[4] (see Sect. 1.3.2). I have termed this Incompatibilism.1. Alternatively, the weight of evidence for a deterministic universe might be considered so strong (and/or the argument for free will so weak) that there is no room left for free will; this would be a "hard deterministic" viewpoint (or, as I have labelled it, Incompatibilism.2, see Sect. 1.3.3).

Since "the free will problem" remains one of the liveliest debates in contemporary philosophy, there have inevitably been many gallons of ink spilt on the topic. For each of these viewpoints (compatibilism, libertarianism, hard determinism) there are multiple nuanced positions, each substantiated (and rebutted) by a variety of closely-reasoned arguments. Within the constraints of the current book it will only be possible to paint descriptions of the major schools of thought on free will and determinism with relatively broad brush strokes, reserving finer detail for

[3]Strictly speaking Chrysippus sought to reconcile a version of 'agent autonomy' with a deterministic universe and made no reference to 'free will' per se. Nevertheless his reflections are held to be one of the earliest attempts at a 'compatibilist' solution to the conundrum.

[4]Of course there is a distinct, political, meaning of Libertarianism. The discussion here is limited to the metaphysical use of the term in the context of free will and moral responsibility.

Table 1.1 Summary of major philosophical positions regarding determinism and free will (Adapted from Roskies 2006, 2012)

		Universe is deterministic?	
		Yes	No
We possess free will?	Yes	Compatibilism (soft determinism)	Libertarianism (Incompatibilism.1)
	No	Hard determinism (Incompatibilism.2)	Random[a]

[a]What is to be made of the fourth sector in Table 1.1, i.e. the view that the universe is nondeterministic *and* that there is no such thing as free will? In one sense this is an admission that everything is random. Roskies (2012) includes an additional classification "Hard incompatibilism" which would span both bottom boxes of the grid. Adherents of this view, she argues, believe free will and moral responsibility are not possible, regardless of whether or not determinism is true (i.e. it would subsume Incompatibilism.2 within a less defined grouping). My view is that this definition adds nothing to those included here and hence it has been omitted

places where greater depth is necessary in consideration of the significance of current developments in genetics and neuroscience.

1.3.1 Compatibilism

As noted above, compatibilism has a long history and remains popular with philosophers. It seems to be the natural way to reconcile our apparent ability to decide between more than one possible action and the revelation, brought to the fore by Isaac Newton, that we live in a world governed by repeatable and predictable laws.

Free will is discussed more fully in the context of Libertarianism below (Sect. 1.3.2). In brief, however, we need to note here that free will has traditionally been seen as having two fundamental characteristics. Firstly, there was a requirement for two or more potential outcomes, with the capacity for the agent to choose between them. In other words there needed to be "alternative possibilities", such that the agent "could have done otherwise". As will be seen, this necessity has been brought into question during more recent discussions. Secondly, there must be an internal source for the decision; responsibility needed to be the agent's choice, not an outcome imposed upon them by external forces (Kane 2011a).

At the risk of gross oversimplification, compatibilists can be divided into classical compatibilists (and their heirs, the conditionalist compatibilists) and new compatibilists (McKenna 2009). For classical compatibilists, who included Thomas Hobbes and David Hume amongst their number, reconciliation between determinism and freedom is achieved by invoking hypothetical or conditional alternative possibilities. This might be summarised in the view "nothing would have prevented you doing differently if you had chosen to do so, but you didn't".

More recently, philosophers have questioned whether the principle of alternate possibilities is necessary in determination of moral responsibility.[5] In a seminal paper of 1969, Harry Frankfurt conducted thought experiments that drove a vital wedge between moral responsibility and the need for alternative possibilities (Frankfurt 1969). Supposing, Frankfurt argues, an agent (Jones) made a wilful decision to carry out a particular action. Before Jones acts, someone else intervenes coercing him into doing the very same act that Jones had already decided to do. When Jones subsequently follows through and carries out this action, is he morally responsible for so doing?

Frankfurt takes us further with a second thought experiment. Here Jones has to decide between Action A and Action B. In this second case, Jones is subject to the authority of a controller, Black. Black has the ability to influence Jones such that, whatever initial decision Jones makes, Black will ensure that *his* choice, Action B, is carried out by Jones. Jones may decide on his own to carry out Action B, in which case Black will do nothing. However, if Jones decides upon Action A, then Black will intervene to ensure that Jones does Action B instead. Frankfurt argues that if Jones decided independently to carry out Action B then he *is* morally responsible, even though he literally could not have done otherwise, since Black would have intervened.

These scenarios may seem contrived, but their logic remains robust and, in consequence, they provided the opportunity for schism between moral responsibility and the necessity for alternative possibilities. In subsequent years, many similar Frankfurt-type thought experiments have been conducted to support this overall position.

1.3.2 *Libertarianism*

A libertarian standpoint must, of necessity, establish positive evidence for the existence of free will, but it is required to do more than this. To be distinct from compatibilism (Sect. 1.3.1), it is also essential for libertarianism to deny determinism and/or to deny that determinism and free will are compatible. Furthermore, the model of free will that emerges from this process must fit in a coherent manner with an indeterministic world.

Before moving on to elaborate arguments in favour of libertarianism, it is important to consider why indeterminism may be warranted. On one level, a simple failure to justify a deterministic view of the world leaves indeterminism in the ascendance. Some advocates of libertarianism, however, maintain that positive evidence in favour of indeterminism can be derived from quantum mechanics (see,

[5]Note that these arguments effectively side-step the notion of free will, jumping instead to the connection between determinism and moral responsibility, the latter being the consequence which would have given relevance to free will itself.

for example, Hodgson 2011). The unpredictability of the behaviour and location of subatomic particles, it is argued, shows us that at a most fundamental level the universe is not determined. Thus, the logic continues, any philosophical position requiring determinism must be incorrect.

Not all commentators are persuaded about the relevance of quantum mechanics to the free will debate. As Randolph Clarke has observed "*Quantum theory is indeed very well confirmed. However, there is nothing approaching a consensus in how to interpret it, on what it shows us with respect to how things are in the world*" (Clarke 2008: p. 19). Some have argued that events at the microphysical level are ineffective in influencing macro-level processes. Furthermore, if quantum level events were shown to have an effect on brain-level processes this would not necessarily offer any endorsement of free will, since uncertainty and freedom are not synonymous.

Robert Kane, one of the most influential libertarian thinkers of the past half century, draws a distinction between "freedom" and "free will". There are, Kane notes, many versions of freedom compatible with determinism. His aim is to substantiate the existence of indeterminist "free will", defined as "*the power to be the ultimate creator and sustainer of one's own ends or purposes*" (Kane 2011b: pp. 382–383).

As discussed above (Sect. 1.3.1), free will is generally held to involve both the existence of alternative possibilities (AP) and identification of processes internal to the agent as being the source of any resulting action. Stressing that *freedom of will* is more than mere *freedom of action*, Kane places greatest emphasis on this second criterion, which he labels "ultimate responsibility" (UR). This is not to render AP as unnecessary for free will. Kane accepts Frankfurt may be correct that for certain decisions there is an absence of AP, however he is insistent that for crucial, character-forming, events in life—events Kane terms "self-forming actions" (SFAs)—then alternatives must have been possible. Not all acts need to be SFAs or "will-setting", the importance of free will does not mean that all actions have to be undetermined, but for those that are, the crucial thing is whether the agent had more than one option which they could choose "*voluntarily, intentionally and rationally*" (Kane 2011b: p. 385). During the course of one's lifetime, an individual will experience an increasing number of SFAs, leading to the formation (and potentially reformation) of character.[6]

Over many years, libertarians have had to face the accusation that a lack of determinism, of necessity, collapses into mere chance. This is a challenge which they feel able to defeat. Imagine a scenario in which an agent is facing conflict between more than one mutually exclusive desire and must make a split-second decision between them (Kane takes the example of a woman on her way to a meeting which will crucially influence her career, who becomes the witness to an ongoing crime and feels a moral duty to intervene).

The critic would say that the agent's decision is a matter of chance, but chance means "out of my control". The libertarian argues that this is not the case. There are indeterministic factors influencing selection between the two, which we can

[6]A criticism made of this model pictures the theoretical scenario in which an agent never faces a genuine SFA in their lives. Does this mean that they are without free will?

perceive as rival neural pathways. Indeterminism might act on those pathways, but it is wrong to think that chance determines which one eventually reaches an activation threshold. If one follows Elizabeth Anscombe and others in picturing indeterministic influences as *hindrances* rather than *causes* of decisions then, for example, an agent might *overcome* the influences of a factor that was leading them to pursue Action B and instead still choose Action A.

What remains crucial is the ability of the agent "*to bring about* whichever *of the options they will,* when *they will do so, for the* reasons *they will do so,* on purpose *rather than by mistake or accident, without being coerced or compelled in doing so, or otherwise controlled by other agents or mechanisms*". (Kane 2011b: p. 389, emphasis in original). The ability that humans have to make such decisions as a result of internal processes, it is argued, distinguish us from brute creature that are tied to responding to the influence of genes and other biological factors (a crucial point to which we will return later, Chap. 5).

1.3.3 Hard Determinism

According to hard determinism, the second version of incompatibilism, the illogicality of free will and determinism co-existing would be revealed via an argument of the following kind (McKenna 2009). (1) An agent acts on his own free will if he could have acted otherwise and/or if he was ultimately responsible for the decision to do so[7]; (2) But if determinism is true it is not possible either to act in other ways or to be the ultimate source of a decision to act in this way; (3) Thus if determinism is true, our agent cannot be acting on the basis of free will.

For the hard determinist, therefore, there is a genuine binary choice; either determinism is real, or free will is real—you cannot legitimately hold on to both. The evidence presented in favour of determinism is frequently based on scientific discoveries. Superficially, this seems to contradict the argument made above, where it was noted that *indeterminists* like to cite developments in science to endorse their position. It is certainly true that the encouragements an 18th or 19th Century determinist[8] would have drawn from the revelations of Newtonian physics have been somewhat undermined by insights coming from quantum mechanics. However, in many other areas of science—genetics, behavioural, cognitive and neuroscience—it is argued that mounting evidence appears to show that our responses are under greater influence of factors outside our conscious control than had previously been imagined.[9]

---

[7]i.e. the traditional criteria for free will which we have seen already in Sect. 1.1.

[8]Though they would more likely have labelled themselves as "necessitarians" than as determinists.

[9]In Sect. 3.2, I will however suggest that the very latest discoveries in biological disciplines are actually questioning once again the degree of determinism in human nature. The point in the current chapter is to rehearse the arguments used by hard determinists not, necessarily, to reflect the views of the present author.

The implications that stem from these scientific discoveries, which we might collectively label biological determinism, are unnerving. As Nobel Prize winner Francis Crick famously stated: "*The Astonishing Hypothesis is that "You", your joys and your sorrows, your memories and your ambitions, your sense of identity and free will, are in fact no more than the behaviour of a vast assembly of nerve cells and their associated molecules. As Lewis Carroll's Alice might have phrased it: "You're nothing but a pack of neurons.*" (Crick 1994: p. 3).

In the context of the present book, it is the consequences of hard determinism for society that offer the greatest challenge. If free will truly is illusory, then notions of moral responsibility, praise and blame are effectively eroded.

References

Beilby, J.K., and P.R. Eddy (eds.). 2002. *Divine foreknowledge: Four views*. Carlisle: Paternoster Press.

Bobzien, S. 2001. *Determinism and freedom in stoic philosophy*. Oxford: Oxford University Press.

Clarke, R. 2008. Incompatibilist (nondeterministic) theories of free will. In *Stanford encyclopedia of philosophy*. http://plato.stanford.edu/entries/incompatibilism-theories. Accessed 25 Aug 2015.

Crick, F. 1994. *The astonishing hypothesis*. London: Simon and Schuster Ltd.

Frankfurt, H.G. 1969. Alternate possibilities and moral responsibility. *The Journal of Philosophy* 66: 829–839.

Hodgson, D. 2011. Quantum physics, consciousness, and free will. In *The Oxford handbook of free will*, 2nd ed, 57–83. New York: Oxford University Press.

Kane, R. 2011a. The contours of contemporary free-will debates (part 2). In *The Oxford handbook of free will*, 2nd ed, 3–35. New York: Oxford University Press.

Kane, R. 2011b. Rethinking free will: new perspectives on an ancient problem. In *The Oxford handbook of free will*, 2nd ed, 381–404. New York: Oxford University Press.

McKenna, M. 2009. Compatibilism. In *Stanford encyclopedia of philosophy*. http://plato.stanford.edu/entries/compatibilism. Accessed 25 Aug 2015.

Monroe, A.E., and B.F. Malle. 2010. From uncaused will to conscious choice: The need to study, not speculate about people's folk concept of free will. *European Review of Philosophy* 1: 211–224.

Nichols, S., and J. Knobe. 2007. Moral responsibility and determinism: the cognitive science of folk intuitions. *Noûs* 41: 663–685.

Roskies, A. 2006. Neuroscientific challenges to free will and responsibility. *Trends in Cognitive Sciences* 10: 419–423.

Roskies, A.L. 2012. How does the neuroscience of decision making bear on our understanding of moral responsibility and free will? *Current Opinion in Neurobiology* 22: 1–5.

Sommers, T. 2010. Experimental philosophy and free will. *Philosophy Compass* 5: 199–212.

Chapter 2
Existing Legislation on Mental Disorders and Criminal Cases

Abstract Having considered philosophical aspects of free will and determinism in the previous chapter, we turn here to reflections on the existing laws regarding mental disorders and criminal responsibility. We will see that legislation and case law in the English-speaking world have resulted in muddled and counter-intuitive verdicts. Rules for determining the sanity of the defendant at the time of the alleged crime are derived from the 19th century M'Naghten case and place emphasis on whether he or she was suffering from a "disease of the mind". It is application of this phrase which has most frequently run contrary to straightforward interpretation. Sensible recommendations for overhaul of the law on criminal liability and mental health have recently been drafted, but are unlikely to be enacted in the foreseeable future.

Keywords Automatism · Criminal liability · Diminished responsibility · Disease of the mind · Insanity · Mens rea · M'Naghten rules

Before reflecting upon the (limited but increasing) uses of biological evidence in criminal cases to date (Chap. 4), and the potential for broader applications of brain science in future cases (Chap. 5), it is important to establish the present legislation regarding mental disorders and criminal responsibility. The emphasis will primarily be on the situation in England and Wales, although the legacy of colonialism means that many countries share a common core of regulation. We will not, here, be considering the growing application of DNA and other forensic evidence relating to the crime itself (*actus reus*). Instead, the focus will be entirely upon the mental or fault-element (*mens rea*) for an offence that has been perpetrated.

Depending upon both the nature of the crime and the alleged cause of mental disorder, there are three defences that might currently be offered; insanity, diminished responsibility and automatism (see Table 2.1). It is important to note from the outset that these prevailing definitions are *legal* not *medical*. This, as will become apparent in subsequent sections, is only one of many problematic aspects in the current application of these rules.

© The Author(s) 2016

C. Willmott, *Biological Determinism, Free Will and Moral Responsibility*, SpringerBriefs in Ethics, DOI 10.1007/978-3-319-30391-8_2

Table 2.1 Summary comparison of insanity, diminished responsibility and automatism (updated[a] from Jefferson 2009: p. 389)

	Insanity	Diminished responsibility	Automatism
Defence to	All offences	Murder	All offences
Cause	Must be internal	Must arise from a recognised medical condition	Must be external
Definition requires	Disease of the mind	Abnormality of mental functioning	Loss of consciousness
Burden of proof	On the accused	On the accused	On the prosecution
Standard of proof	Balance of probabilities	Balance of probabilities	Beyond reasonable doubt
Outcome if plea successful	Not guilty by reason of insanity (special verdict)	(Voluntary) manslaughter	Acquittal

[a]Criteria for a plea of diminished responsibility were originally described in the Homicide Act 1957 s2. This was amended by the Coroners and Justice Act 2009 s52

2.1 Insanity

Someone might be deemed insane at the time of their trial, and therefore "unfit to plead" (Law Commission 2010).[1] Alternatively, they might be considered sane at the time of the trial, but with one or more sides in the case arguing that they were insane at the time the crime was committed.[2] It is this latter scenario, in which the so called "special verdict" of "not guilty by reason of insanity" would be given, that is of greater interest in the current context.

In some senses, insanity is both the best place and the worst place to start a consideration of legal approaches to mental disorders and culpability. It is the best place since many of the rules governing the definition of insanity are essentially unaltered since the M'Naghten case of 1843.[3] It is the worst place because it is only infrequently used as a defence in the UK, fewer than five times per year throughout most of the 1970s and 1980s (Jefferson 2009). Following the Criminal Procedures (Insanity and Unfitness to Plead) Act 1991, this number has risen to 20–30 special verdicts per year; still a small drop in the ocean of about 90,000 people per annum tried in the Crown Court (Law Commission 2012). It is also a poor place to start

[1]The evolution of "unfitness to plead" is nicely summarised in a consultation paper produced by the Law Commission 2010.

[2]One high profile example of the sanity of the defendant was questioned occurred during the 2012 trial in Norway of Anders Breivik who was responsible for the murder of 77 people in the summer of 2011. Interestingly it was the prosecution on this occasion that had sought to show he was insane; Breivik himself had wished to be declared sane to support the argument that his actions were politically motivated and not the work of a deluded mind.

[3]M'Naghten's Case 8 ER 718, [1843] UKHL J16.

since the rules outlined by the presiding judge Lord Tindal CJ (see Sect. 2.1.1) have been interpreted in contradictory ways during the subsequent 170 years.

2.1.1 *The M'Naghten Case*

In 1843, Scottish woodturner Daniel M'Naghten was put on trial for the murder of Edward Drummond, but was found not guilty by reason of insanity. The judgment might have passed unnoticed except that the intended target was not Drummond, but rather his boss, Prime Minister Robert Peel. Not for the last time, the metaphorical proximity of a national political leader to the crime led to post hoc questioning of the ruling.

The matter was referred to the House of Lords and a debate on the concept of insanity. During the discussion, Lord Tindal observed that: "*the jurors ought to be told in all cases that every man is to be presumed to be sane, and to possess a sufficient degree of reason to be responsible for his crimes, until the contrary be proved to their satisfaction; and that to establish a defence on the ground of insanity, it must be clearly proved that, at the time of the committing of the act, the party accused was labouring under such a defect of reason, from disease of the mind, as not to know the nature and quality of the act he was doing; or, if he did know it, that he did not know he was doing what was wrong*." (quoted by Clarkson et al. 2010: p. 378).

There are several relevant points to note. Firstly, there is the presumption of sanity and, in consequence, responsibility. Secondly, insanity is stated as originating with "disease of the mind", which thirdly, causes "defect of reason" at the time of the crime. Fourthly, this defect of reason is manifest as a lack of discernment regarding either the nature and quality of the act, or that it was wrong. Because of this emphasis on knowing right from wrong, this has subsequently been referred to as a "cognitive test" of criminal responsibility (in contrast to a "volitional test", which we will mention later).

2.1.2 *Interpretation of the M'Naghten Rules*

Pivotal within the M'Naghten rules is the notion of "disease of the mind". It is this phrase, more than any other, which has proven problematic in the application of the rules both in the UK and overseas. It also has implications when we move on to consider the potential relevance of emerging biological understanding about the workings of the brain (Chap. 3).

Case law has shown that the "disease of the mind" need not be a mental illness, nor indeed a disease at all in a strict medical sense. Probably the most important demarcation to emerge has been the distinction between "internal causes" and "external causes". For a defence of insanity, the cause must be internal; if the cause is external then it is considered a case of automatism (Sect. 2.3). This distinction has led to a series of judgments which, when seen side by side, look most peculiar; none more so than in regard to diabetic defendants.

A diabetic who was involved in a criminal incident as a consequence of going hypoglycaemic, that is they had excessively low blood sugar, was found to be exhibiting automatism (because the condition is caused by taking insulin, which was deemed to be an external cause),[4] whereas another diabetic who went hyperglycaemic (i.e. they have excessively high blood sugar, the natural outcome if they have not taken insulin) was found to be insane.[5] As Clarkson and colleagues note "*we are left with a law under which some diabetics will be able to secure a complete acquittal while others will be regarded as insane. Such a position is absurd*" (Clarkson et al. 2010: p. 383). At other times epilepsy,[6] atherosclerosis (hardening of the arteries)[7] and sleepwalking[8] have all been classified as diseases of the mind leading on to the special verdict.

Anomalies of these kinds have been offered, over many years, as evidence in calls to reform legislation in this area (see Sect. 2.4: Reform of the law). In 2012, the UK Law Commission conducted a consultation on Insanity and Automatism (Law Commission 2012), with their recommendations reported the following year (Law Commission 2013a). Their proposals are outlined in Sect. 2.4 (where it will also be explained why these are unlikely to be enacted in the near future). As it stands, therefore, the definition of insanity remains largely unaltered from M'Naghten.

There have, nonetheless, been some developments in the potential consequences of being found "not guilty by reason of insanity". As originally drafted, a successful defence of insanity would have led to detention in a secure hospital, possibly "with restrictions", i.e. until the Home Secretary deemed it appropriate for them to be released. This is clearly inappropriate for a diabetic who failed to remember to take their insulin, or an epileptic who causes harm whilst having a seizure. If the crime was murder, then hospitalisation for life was mandatory.

Since the Criminal Procedures (Insanity and Unfitness to Plead) Act 1991 came into force (in 1992) and was subsequently amended by the Domestic Violence, Crime and Victims Act 2004, judges have been granted more discretion regarding sentencing. A judge can now decide on the basis of medical evidence from two doctors, whether she considers the accused is unfit to plead. If, after this assessment, the judge *does* elect to go forward to trial, then a "trial of the facts" ensues, in which the prosecution must convince a jury beyond reasonable doubt that the defendant did carry out the crime for which they stand accused. If the jury find against the

---

[4]Quick [1973] QB 910.

[5]Hennessy [1989] 1 WLR 287.

[6]Sullivan [1984] AC 156.

[7]Kemp [1957] 1 QB 399.

[8]Burgess [1991] 2 QB 92, though there is more recent evidence that somnambulism is more likely now to be treated as a case of automatism.

defendant then the judge can choose from a range of sentences from detention in a psychiatric hospital, with restrictions, down to an absolute discharge.

2.2 Diminished Responsibility

Without being as susceptible to contradictory interpretations as the rules on insanity, there remain a number of complications regarding the defence of diminished responsibility. Most significantly, diminished responsibility is only applicable in cases of murder (or assisting in murder, but *not* attempted murder).[9]

Historically, this was appropriate; the defence stems from an era when the punishment for murder would have been execution.[10] It is really a *partial* defence; pleading diminished responsibility was de facto an admission to having killed someone, but offered an avenue for reducing the sentence to, for example, life imprisonment in circumstances where some explanation for the actions of the defendant might be offered. Since we are no longer in the habit of applying death sentences, at least not in the UK, some have argued that diminished responsibility should either be broadened to cover other crimes as well or abandoned as a defence (particularly if the punishment for murder is reduced from the current mandatory life sentence). In the chess game of legislation, claiming diminished responsibility is not without its risks. A successful appeal on these grounds leads inevitably to being found guilty of manslaughter and may still result in a long period of incarceration.

The original description of a person suffering from diminished responsibility in the Homicide Act 1957 required that he exhibited "*substantially impaired... mental responsibility*" caused by "*abnormality of mind*".[11] As a consequence of the Coroners and Justice Act 2009, this has been changed to "*abnormality of mental functioning*". The cause of this abnormality must be a recognised medical condition which substantially impaired the defendant's ability to understand the nature of their conduct, to form a rational judgement, and/or to exercise self-control.[12]

2.3 Automatism

Automatism has been defined as "*an involuntary movement of the body or limbs of a person [following] a complete destruction of voluntary control*".[13] A successful plea of automatism must demonstrate that at the time of the alleged offence, the

[9]The defence of diminished responsibility was introduced via the Homicide Act 1957.

[10]This, of course, remains a potential sentence in the majority of US states (see Chap. 4).

[11]Homicide Act 1957 s2.

[12]Coroners and Justice Act 2009 s52.

[13]Winn J, in Watmore v Jenkins [1962] 2 QB 572, 587.

accused was out of control as a consequence of an "external" influence, not something that they might reasonably been expected to predict. A finding in favour of the defendant leads to an outright acquittal.

A well-known case has bearing for the later debate regarding the current understanding of neuroscience. A man, Charlson, invited his son to look out of the window at a rat in the river below, whereupon he hit his son on the head with a mallet and threw him out of the window.[14] This may look superficially like a clear-cut case of insanity, but it was decided that he was suffering from automatism when it was revealed that the man had a brain tumour.

This verdict has been controversial on several grounds. Firstly, it clearly involves a physical abnormality within the man's brain, but was considered an "external" influence and not a "disease of the mind". Secondly, the man had committed a violent assault yet effectively walked free, despite the fact he may have remained a danger to others.

The case is therefore a reminder that there may be several conflicting reasons for a decision regarding the fate of a defendant. It has variously been suggested that sentencing can serve as a means towards retribution against the defendant, to act as a deterrent to others or to protect society (Eastman and Campbell 2006). This latter reason is the most prominent in cases of hospitalisation in a secure unit, alongside potential treatment and rehabilitation of the defendant.

2.4 Reform of the Law

As noted above, the Law Commission recently published their recommendations for revision of English Law concerning criminal liability (Law Commission 2013a). The specifics of their proposals will be outlined below (Sect. 2.4.1). Connoisseurs of the history of this field will not, however, be holding their breath whilst they await implementation of any changes. Erudite suggestions for reform have been made on many occasions but the majority have failed to make it onto the statute books.

Criticism of the M'Naghten rules go back almost as far as the rules themselves. Public expressions of dissatisfaction have been more pronounced since 1953, when the Royal Commission on Capital Punishment condemned the "*manifest absurdity of the M'Naghten test*" and recommended that the rules be abolished.[15] This did not happen, but the Commission did lead onto the Homicide Act 1957 and the introduction of the defence of diminished responsibility. A second proposed defence—that of "irresistible impulse"—was not included in the Act (Jefferson 2009).

---

[14]Charlson [1955] 1 All ER 859.

[15]Royal Commission on Capital Punishment, Cmnd 8932 (1953), p 104.

The Report of the Committee on Mentally Abnormal Offenders 1975, usually referred to as the Butler Committee,[16] proposed a new verdict of "not guilty on evidence of mental disorder". They also sought to define a mental disorder in a much more natural way than the M'Naghten rules suggested. Regarding issues that were clearly brain-related (lasting impairment of intellectual function, lasting alteration of mood, delusional beliefs, delusional misinterpretation of events) there would be a broadening of the definition, but the anomalies thrown up by case law (diabetes, epilepsy, sleepwalking) would not be included.

The Butler Committee recommendations have informed discussion, and been widely praised, in several subsequent reports, but the majority of changes have not been implemented. Some changes, such as diversification of the sentencing options after a special verdict *have* been enacted (Sect. 2.1), but at the core the M'Naghten rules still carry undue importance in English courts.

Over recent years, other jurisdictions in the English-speaking world have acted to amend their laws on mental disorders and criminal responsibility. These include Canada (1985),[17] Ireland (2006),[18] and Scotland (2010).[19] Here is not the place to discuss each of these in detail. Before returning to look at recent developments in England, and in light of the prominence that American cases will play in Chap. 4, it is however appropriate to pause briefly in order to elaborate on the evolution of the insanity defence in the United States of America, and particularly on the impact of the trail of John Hinckley upon the evolution of legislation in the USA.

In common with the UK, the laws governing criminal responsibility and the insanity defence have their roots in the M'Naghten case. At various times, as far back as 1887,[20] there have been concerns that the definition of insanity offered by the M'Naghten test was too narrow. The emphasis on the cognitive dimension, meant the potential criminalisation, it was alleged, of individuals who *knew* what they were doing was wrong but, for whatever reason, were unable to stop themselves acting on the impulse to do wrong. There was therefore calls for inclusion of a volitional test; to broaden the definition of insanity to cover those individuals for whom "disease of the mind" had impaired their ability to resist an action rather than knowledge that to do so would be wrong.

The volitional and the cognitive dimensions of insanity defence were formally brought together in the American Law Institute's Model Penal Code (MPC) of 1962. The MPC stated "*A person is not responsible for criminal conduct if at the time of such conduct as a result of mental disease or defect he lacks substantial capacity either to appreciate the criminality [wrongfulness] of his conduct or to*

[16]Report of the Committee on Mentally Abnormal Offenders (Butler) Cmnd 6244 (1975).

[17]Criminal Code, R.S.C., 1985, c. C-46, as amended.

[18]Criminal Law (Insanity) Act 2006.

[19]Criminal Justice and Licensing (Scotland) Act 2010.

[20]Parsons v State, 2So. 854 (Alabama 1887).

conform his conduct to the requirements of the law".[21] These standards were adopted by the DC Court of Appeal in 1972.[22]

Several aspects of the MPC prompted opposition. In particular, there were concerns that the prosecution were called upon to prove beyond reasonable doubt that the defendant was not insane. Coupled with the lack of a test to unambiguously distinguish an *inability* to avoid an action from an *unwillingness* to do so. These concerns came into sharp relief with the case of John Hinckley.[23]

In March 1981, Hinckley attempted to assassinate the American President Ronald Reagan. At his trial the following year, Hinckley was found not guilty by reason of insanity (the attack had been his attempt to impress the actress Jodie Foster with whom he was besotted).

The Hinckley verdict caused widespread outrage in the USA. In consequence, Congress, and half of the individual states tighten up their rules, with many becoming much closer to the M'Naghten rules than had previously been the case.

2.4.1 2013 Law Commission Discussion Paper

The discussion paper Criminal Liability: Insanity and Automatism (Law Commission 2013a) represents a significant attempt to chart an appropriate course for revision of the law in England regarding mental disorders and criminal responsibility. Unsurprisingly, the authors were scathing about the inadequacies of the current regulations. They propose abolition of the defence of insanity, and significant alteration to the defence of automatism in the light of other changes. In their place, the main defence would become "not criminally responsible by reason of recognised medical condition".

There are several points to note regarding the phraseology of the new defence. Firstly, the notion of "guilt" is replaced by "criminal responsibility" to fit an ambition to ensure that people are only punished on grounds for which they are appropriately accountable. Secondly, emphasis on "recognised medical conditions" tidies up the anomalies that have arisen from case law interpretation of "disease of the mind". Indeed, the new wording would broaden the definition of underlying causes to include physical as well as mental conditions, and would naturally encompass cases where loss of control arose from diabetes, epilepsy and sleep disorders. Thirdly, "recognised medical conditions" makes it immediately apparent that we are now working with appropriate medical definitions, not quasi-medical legal definitions, of disease. It also builds in provision for evolution of our understanding of the underlying causes of conditions.

---

[21]Model Penal Code (Proposed Official Draft 1962).

[22]United States v Brawner, 471 F.2d 969 (D.C. Cir. 1972).

[23]United States v Hinckley, 672 F.2d 115 (D.C. Cir. 1982).

If this change was brought in, there would be a corresponding re-definition of the defence of automatism. The latter would now be restricted to situations in which there was a "total loss of capacity to control one's actions which is not caused by a recognised medical condition" (Law Commission 2013b, para 118). Grey areas arising from the definition of internal factors v external factors would thus be removed.

The authors of the Law Commission Discussion Paper emphasise that existing medical conditions per se do not necessarily provide automatic exemption from prosecution. For example, if someone with a recognised sleeping disorder caused an accident whilst driving, having chosen to ignore tell-tale signs of their increasing drowsiness, then they exhibit "prior fault". Similarly, loss of capacity arising from voluntary intoxication would not be legitimate defence.

Overall, the Law Commission recommendations are logical, thorough and a vast improvement on the present legislations. As with earlier recommendations, however, it remains highly unlikely that these new proposals will be enacted any time soon.

History has shown that changes can come speedily when politicians recognise some vested interest. The original establishment of the M'Naghten rules in the UK was motivated, at least in part, by the fact that the Prime Minister was the intended target. In the USA, the fact that the President was the victim prompted the tightening of insanity rules following the acquittal of John Hinckley.

At other times the lack of political will to give a Bill the necessary time in parliament has seen potentially valuable amendments flounder. Authors of Criminal Law textbooks (e.g. Clarkson et al. 2010; Jefferson 2009) write wistfully of opportunities missed with the failure to see enacted the suggestions of various reports, particularly the Butler Committee 1975 and the Draft Criminal Code (Law Commission 1989).

In the case of the 2013 Law Commission Discussion Paper, the authors themselves recognise that there are actually more pressing changes within English Law that need to be made. In particular, they point to concerns regarding miscarriages of justice arising from a defendant's mental health at the time of their trial.

Despite, or perhaps because of, the obvious flaws in legislation on insanity and automatism when a crime is perpetrated, lawyers have developed effective work arounds that limit harms potentially arising due to these idiosyncrasies. Difficulties in the fair treatment of defendants who are, or should have been, identified as "unfit to plead" at trial are seen as a priority because they may affect a larger number of people and would 'filter down' to influence cases where the mental well-being of a defendant at both the time of the crime and the time of the trial are in doubt.

It is interesting to note, however, that should the changes recommended in the Law Commission discussion paper on Insanity and Automatism be enacted, the emphasis on a "recognised medical condition" might create room for greater consideration of the role played by deeper knowledge of biological factors in criminal behaviour. Before returning to that issue in Chaps. 4 and 5, let us first move on to consider some of the emerging data on genetic and neuroscientific aspects of behaviour.

References

Clarkson, C.M.V., H.M. Keating, and S.R. Cunningham. 2010. *Clarkson and Keating criminal law: Text and materials*, 7th ed. London, UK: Sweet and Maxwell.

Eastman, N., and C. Campbell. 2006. Neuroscience and legal determination of criminal responsibility. *Nature Reviews Neuroscience* 7: 311–318.

Jefferson, M. 2009. *Criminal law*, 9th ed. Harlow, UK: Pearson Education Ltd.

Law Commission. 1989. A criminal code for England and Wales, Report 177. Available online at http://www.bailii.org/ew/other/EWLC/1989/177_1.pdf. Last accessed 21st Aug 2014.

Law Commission. 2010. "Unfitness to plead", consultation Paper 197. Available online at http://lawcommission.justice.gov.uk/docs/cp197_Unfitness_to_Plead_consultation.pdf. Last accessed 21st Aug 2014.

Law Commission. 2012. Insanity and automatism: A scoping paper. Available online at http://lawcommission.justice.gov.uk/docs/insanity_scoping.pdf. Last accessed 21st Aug 2014.

Law Commission. 2013a. Criminal liability: Insanity and automatism—a discussion paper. Available online at http://lawcommission.justice.gov.uk/docs/insanity_discussion.pdf. Last accessed 21st Aug 2014.

Law Commission. 2013b. Criminal liability: Insanity and automatism—a discussion paper: summary for non-specialists. Available online at http://lawcommission.justice.gov.uk/docs/insanity_discussion_summary.pdf. Last accessed 21st Aug 2014.

Chapter 3
Biological Basis of Behaviour

Abstract With the exception of the most involuntary of reflexes, my actions are the consequence of the processes going on within my brain. In this chapter, our focus switches to scientific approaches which offer insight into the underlying biology of behaviour. Findings emerging from recent developments in both genetics and neuroscience are considered. The strengths and weaknesses of a range of methodologies used for imaging the brain are reviewed, with particular attention given to the roles played by Electroencephalography, Positron Emission Tomography and functional Magnetic Resonance Imaging in the study of operational brains. We reflect on the implications for "free will" of a series of experiments which appear to demonstrate brain activity prior to conscious awareness of decision. Validated examples of genes involved in behaviour are discussed, alongside the growing importance of an understanding of the molecular mechanism by which expression of "hard wired" genes can be moderated by environmental stimuli.

Keywords Behavioural genetics · Brain imaging · EEG · Epigenetics · fMRI · G × E interactions · MAOA · P300 · Readiness potential · Serotonin

It the preceding chapters we have considered both philosophical and legal dimensions to responsibility. It is against this backdrop that current research into the biological basis of behaviour must be considered.

The potential implications of any brain imaging or genetic data which countermanded the traditional emphasis on free will are profound. If evidence could be found that behaviour was in some way determined then, regardless of the nuances of compatibilism discussed in Chap. 1, there might be a significant impact on the legal system. As Adina Roskies notes "*We care about free will primarily because we care about what comes along with it—moral responsibility.*" (Roskies 2006: p 421).

So what is the evidence regarding the "hard-wiring" of human conduct? To address this issue we will reflect in turn on research into the structural and functional architecture of the brain, and into the genetics of behaviour.

© The Author(s) 2016

C. Willmott, *Biological Determinism, Free Will and Moral Responsibility*, SpringerBriefs in Ethics, DOI 10.1007/978-3-319-30391-8_3

3.1 Neuroarchitecture and Brain Imaging

The pivotal role played by the brain in so many aspects of life, not least in moral decision making, has made it the subject of much curiosity. Of course, the fact that the brain is housed inside a hard cranium, specifically designed to protect it, does present a fundamental difficulty when it comes to studying how it works. Attempts to map out the link between the structure of the brain and its various functions have often involved anatomists cutting through the skull in order to get at the tissue inside. This is clearly an invasive and risky procedure. For this reason the majority of human anatomic work has been conducted post mortem, though de facto this has limited the capacity to learn lessons about the operational brain. Scientists and clinicians have therefore looked for various ways to map the brain without having to carry out such literal dissection. Before moving on to examine some of the implications emerging from brain analysis, it is appropriate to outline both the techniques that have been used to view inside people's heads and some of the emerging knowledge about the functions performed by different regions of the brain.

3.1.1 Brain Imaging Techniques

Given the explosion of new imaging technologies in the past fifty years (Table 3.1), it would be easy to overlook the importance that has been played by a far more establish methodology, the X-ray.

First developed in the 1890s, X-rays made an almost immediate impact, finding use in both medicinal and legal contexts before the close of the 19th Century (Brogdon and Lichtenstein 1998). The first criminal case employing X-rays actually involved visualisation of someone's leg rather than a brain; George Holder of

Table 3.1 A summary of brain imaging techniques and decades during which they were first used in a clinical and/or a legal context

Technique	First used
X-rays	1890s
➥ Computer Tomography (CT scan)[a]	1970s
➥ Single-photon emission computed tomography (SPECT)	1990s
Electroencephalography (EEG)	1920s
➥ Brain Fingerprinting	1980s
Positron Emission Tomography (PET)	1960s
Magnetic Resonance Imaging (MRI)	1970s
➥ functional Magnetic Resonance Imaging (fMRI)	1990s

[a]The symbol ➥ is used to indicate family relationships, where one technology has evolved from another

Montreal was convicted of attempted murder after an X-ray confirmed the existence of a bullet in the lower limb of Tolson Cunning.

When it comes to study of the brain, imaging techniques can be divided into those that reveal structural information and those that purport to show correspondence between specific functions and particular subsection(s) of the organ. X-rays, Computed Tomography (CT or CAT scans) and Magnetic Resonance Imaging (MRI) fall into the former category, whereas Single-photon emission computed tomography (SPECT), Electroencephalography (EEG), Positron Emission Tomography (PET), functional Magnetic Resonance Imaging (fMRI) and Brain Fingerprinting report on the operation of the brain. Attempts have been made to introduce each of these approaches as evidence in court, with mixed rates of success (see Sect. 4.4).

Structural imaging approaches: **X-rays** can reveal information about the internal structure of the body because materials of different composition, density and/or thickness absorb electromagnetic radiation with varying efficiency. The X-rays passing through the body can be captured on appropriately sensitive film to produce a two-dimensional image.

In truth, conventional X-rays reveal little about the detailed structure of the brain. However **CT** scanning, whilst employing the same fundamental physics, improves considerably upon basic X-ray imaging by developing 3D representations of the brain. With the help of computers, a series of images can be produced as 'slices', which can then be reconstructed into a 3D view.

Unlike X-rays and CT scans, which involve the use of potentially harmful ionising radiation, **MRI** exploits the inherent magnetic properties of atoms, specifically hydrogen atoms due to their abundance in fats and water, which constitutes the majority of the human body. The atoms would naturally spin in random orientations. A patient (or other research subject) lies flat within the MRI machine, surrounded by a strong magnet. When it is switched on, the magnet causes all of the hydrogen atoms to spin in a coordinated direction, aligned head-to-toe, or vice versa (Berger 2002).

Pulses of radio waves at a specific frequency appropriate to disrupt the spin of hydrogen atoms are then directed to the relevant section of the body, in this case the head. Energy from the radio waves causes some of the hydrogen atoms to adopt a different orientation. When the radio signal is turned off, the hydrogen atoms return to their original orientation, as determined by the magnetic field. This process is known as "relaxation". Different tissues, including tumours, allow relaxation to occur at different rates and/or may contain different percentages of water. With the aid of a computer, mathematical information about relaxation can be converted into three-dimensional brain images.

As well as the safety issues noted above, MRI has certain advantages over X-rays and CT scans. In particular, MRI is better than X-ray based systems at visualising soft tissues and gives sharper images. There are downsides though; MRI machines are noisy and claustrophobic, and very expensive.

Functional imaging approaches: Both PET and SPECT require the use of radioactive isotopes to report on metabolic activity within the brain. In **PET**, radioactive decay causes the release of a positron, a positively charged particle. When a positron encounters an electron in the tissue under investigation, the two particles interact and "cancel each other out" leading to the generation of two gamma-rays. These rays are detected using scintillant material in the walls of the machine surrounding the patient's head. The scintillant in turn emits light which is analysed by computer.

The radio-labelled compound, typically a glucose analog fludeoxyglucose (^{18}F-FDG) or labelled water ($H_2{}^{15}O$), is delivered intravenously. The level of radioactivity seen in different parts of the brain is a reflection of the blood flow and/or glucose metabolism in those regions which, in turn, is taken to be a proxy for brain activity.

For PET, the radioisotopes are short lived; ^{18}F has a half-life of 110 min, ^{15}O just 2 min (Muehllehner and Karp 2006). In consequence, facilities utilising this technique must be geographically close to a laboratory producing the labelled metabolite, contributing to the uneven availability of this service. For **SPECT**, isotopes such as ^{99m}Technetium and ^{201}Thallium are used. The radioactive element do not naturally occur within biomolecules, so radiopharmaceuticals such as ^{99m}Tc-hexamethylpropyleneamine oxime must be specially synthesised (Ollinger and Fessler 1997). These isotopes used for SPECT do however have longer half-lives than those used for PET meaning that machines capable of obtaining SPECT data are more readily accessible.

Like PET, SPECT involves production of gamma-rays, but despite use of a rotating gamma camera, SPECT scans result in lower resolution images that PET. Nowadays both SPECT and PET can be conducted in combination with traditional CT scans to generate functional and structural data at the same time.

The use of **fMRI** represents an important advancement in brain mapping because it does not require the use of radioactivity. Instead fMRI can be used to map active regions of the brain by exploiting natural differences in the magnetic properties of haemoglobin in oxygenated blood versus deoxygenated blood. By comparing images of brains at rest versus brains engaged in specific tasks, a blood oxygen level dependent (BOLD) contrast can be generated. In this way it becomes possible to visualise the areas of the brain requiring greater supply of oxygen, which is taken to represent brain activity.

Waveform approaches: EEG and Brain Fingerprinting differ from the methods discussed thus far. Rather than an image of the brain, these methods output measurements as a series of waves representing brain activity, known as a "montage". **EEG** measures electrical activity in the brain arising from the coordinated response of thousands of neurons. It involves recording signals at multiple electrodes positioned on the scalp.

Compared with most of the other imaging techniques, EEG offers worse spatial accuracy, but clearer temporal resolution. It also requires less specialist equipment that many of the scan-based systems and, in consequence, is both cheaper to use and more flexible regarding location (e.g. it can genuinely be used for "bed-side" analysis in a way that fMRI cannot).

Brain Fingerprinting is a particular variation of EEG seeking to compare brain responses to three kinds of stimuli (Farwell and Smith 2001). The first type of stimulus involves particular things that the subject had been asked to memorise. These are known as "targets" and serve as positive controls for the process. Secondly, there are stimuli with which the subject is not expected to have any connection. These are the "irrelevants" and are negative controls. Thirdly, there are the stimuli under test, the "probes"—things which the subject may know, for example, as a consequence of their presence at a crime scene. Such "guilty knowledge" might be considered to be incriminating (see Sect. 4.4.1).

Scientists employing this method look for evidence of Memory and Encoding-Related Multifacted Electroencephalographic Responses (MERMERs). The most important MERMER is known as the P300 wave, an involuntary "event-related potential" (ERP) occurring some 300 ms after the trigger stimulus.

3.1.2 *Functional Architecture of the Brain*

The human brain is divided into six main parts; the midbrain, the pons and the medulla oblongata (these three collectively constituting the brain stem), the cerebellum, the diencephalon (which includes the thalamus and the hypothalamus), and the cerebrum (Kandel et al. 2000). During the early part of the 20th Century, painstaking anatomical work was conducted by German anatomist Korbinian Brodmann. By examining variation in the organisation of different cell types, Brodmann defined 52 structurally-distinct areas within the human cerebral cortex (Kandel et al. 2000). Although more modern techniques have brought into question some of his original mapping of functions to particular regions of the brain, Brodmann's cytoarchitectonic numbering of areas has nonetheless served for several decades as a valuable starting point for defining the substructure of the brain.

The cerebral hemispheres can initially be divided into four distinct lobes; occipital, parietal, temporal and frontal. As the name suggests, the frontal lobe is situated in the anterior part of the brain, i.e. on the facial side of the head. The frontal lobe is proportionately larger in primates than other mammals, and all the more so in humans (Miller and Cohen 2001). It has a pivotal role in higher brain functions, i.e. not the innate, routine responses maintained by other regions. Several subsections within the frontal lobe have been closely identified with specific behaviours (see Table 3.2). In particular, the PreFrontal Cortex (PFC) and divisions thereof have crucial roles to play in cognitive control or "executive functions", which can be defined as "*those high-level processes that control and organise other mental processes*" (Gilbert and Burgess 2008: p. R112).

The PFC has been shown to play a crucial role in "working memory", i.e. retention of short-term information required to fulfil a goal, and in "behavioural inhibition", which includes the active exclusion of information irrelevant to completion of the task and the wilful decision not to go through with an action, such as giving into an inappropriate temptation (Bunge et al. 2001; Gilbert and Burgess

Table 3.2 Correlation of certain behaviours with subsections of the frontal lobe of the human brain

Area of brain	Associated behaviour-related activity
Anterior Cingulated Cortex (ACC)	Inhibitory control, emotional processing (including empathy), detection of cognitive conflict
Orbital PreFrontal Cortex (OPFC)	Decision-making, emotional processing (including regret)
Ventromedial PFC (VMPFC)	Ethical decision-making, fear and processing of risk
Ventrolateral PFC (VLPFC)	Inhibition of behaviour
Dorsolateral PFC (DLPFC)	Reasoning, behavioural control, cognitive flexibility, impulse control
Amygdala	Emotional learning and memory, auditory and facial emotion recognition, fear conditioning
Supplementary Motor Complex (SMC)	Volitional (self-initiated) movements, inhibition of action and response to alteration in planned activity

Original concept for table inspired by Mobbs et al. (2007). Additional information drawn from: Balleine and Killcross (2006), Brierley et al. (2002), Bunge et al. (2001), Gilbert and Burgess (2008), Nachev et al. (2008), and Salat et al. (2002)

2008). Any notion of free will and "top-down control" will, of necessity, involve influences exerted via the PFC.

3.2 Genetics of Behaviour

We are going to move on to think about the potential role of specific genes in human behaviour. Before doing so, however, it should be noted that psychologists and other behavioural scientists have employed a variety of alternative approaches which give strong support to the idea that at least some of our behaviour stems from our genetic make-up.

As far back as the 1920s, scientists have sought to identify a potential role for hereditable factors by comparing the extent to which characteristics are shared by "identical" or monozygotic (MZ) twins in contrast to "fraternal" or dizygotic (DZ) twins. The premise, which is known to have potentially confounding limitations, is that all twins (MZ and DZ) will have had a shared environment (both in utero and after birth). However only MZ arise from the same fertilised egg and will have "identical" genetic heritage; DZ twins are no more genetically alike than other siblings. Thus traits found more commonly in MZ than DZ twins may be attributable to genetics.

As Andrews and Bonta (2010) have observed, the perfect opportunity for analysis would come if MZ twins were separated at birth and raised entirely separately in different families. For obvious reasons it would be unethical to conduct such an intervention purely to further our knowledge of genetics. However there are rare situations when, for other reasons, identical twins have been separated early in life and raised apart in adopted families.

This does herald the second non-molecular way the influence of genetics on behaviour has been investigated, which is the study of adoptees. Antisocial behaviour, for example, exhibited by an adoptee can be compared against the equivalent characteristic in his biological parents, his adoptive parents and/or genetically unrelated siblings in his adoptive family. This will help to delineate genetic versus environmental factors.

Via these various mechanisms, it has been possible to establish with some confidence that criminality and/or other antisocial behaviour is attributable, at least in part, to genetic factors. Calculations of the genetic contribution to behaviour vary significantly (not least due to methodological differences), however it is not considered unreasonable to attribute approximately 30 % of antisocial behaviour to genetic factors (see Rhee and Waldman 2002; Andrews and Bonta 2010 for more detailed analysis).

Of course it is important to remember that studies of these types do not offer information about the specific contribution of genes to the criminal behaviour of any particular individual, such as the defendant in a murder trial (a scenario to which we will return in Chap. 4). A third approach would be to interrogate the pedigree of a person, for example our hypothetical defendant, to examine the inheritance of a feature of interest through the generations in their family tree. If undertaken appropriately, a study of this kind will likely generate a "genogram", a pictorial representation of the relationships within the family with emphasis on the patterns of inheritance of the trait(s) under consideration. In the present context this could be a history of criminality within the family, but it might be somewhat more subtle than this, looking at family members with a history of mental illness, alcoholism or similar.

From our current vantage point, in which the complete genome[1] of an individual can be sequenced in a few days for low thousands of dollars, it is becoming possible to add a molecular dimension to these more traditional approaches. It is to the potential for genetic analysis that we now turn.

Discovery of the structure of deoxyribonucleic acid (DNA), and the subsequent elucidation of the ways in which proteins are encoded by that DNA, opens up the potential for interrogation of the underlying molecular biology of behaviour. The identification of mutations within that coding sequence as the cause of various inheritable diseases has driven a view amongst some scientists that this "book of life" is all that we need. It is no surprise, for example, that it was Francis Crick, one of the authors of the seminal paper on the structure of DNA (Watson and Crick 1953), who made the bold assertion quoted in Sect. 1.3.3, regarding the

[1]A distinction is drawn between "genomics" in which a significant number of different genes are investigated at the same time (possibly the entire genetic make-up of an individual) as against "genetics" where the focus is more likely to be on relationship with one or two specific genes.

deterministic character of nature [Crick's co-author, James Watson, is not quite as forthright, but shares some of the same sentiment when he observes that "*In large measure, our fate is in our genes*" (quoted by Alper 1998)].

This reductionist[2] model of life is not uniformly endorsed (e.g. see Noble 2006, for a well-reasoned critique). A good argument can be made that many of the most recent discoveries in genetics are actually demonstrating that this view was overly simplistic. Evidence from a diverse range of biological disciplines, from botany to psychology, has progressively seen models involving Gene-environment (G × E) interactions replacing naïve notions of Nature versus Nurture (Baum 2013). Not only does data increasingly point to "both-and" explanations rather than "either-or", but scientists are also beginning to understand the molecular mechanisms by which environmental factors can exercise influence on gene expression. These mechanisms include both epigenetics (Sect. 3.2.1) via chromatin modification and the influence of non-coding RNAs (Sect. 3.2.2).

3.2.1 Epigenetics

Over the past fifteen years, there has been a paradigm shift in the understanding of gene expression. In particular, there has been recognition that mechanisms exist for the transmission of inheritable changes in DNA expression that do not involve mutation of the DNA coding sequence per se. This phenomenon, known as epigenetics, has been defined as "*the structural adaptation of chromosomal regions so as to register, signal or perpetuate altered activity states*" (Bird 2007: p 398). Epigenetic alterations in gene expression are generally achieved by the attachment (or removal) of a range of small molecules to the DNA and/or to the proteins within the histone complexes, around which the nucleic acid is wrapped. Methylation of DNA at so-called "CpG Islands" located close to the transcription start site for a gene (i.e. the position where the "message" begins) can influence whether or not it is expressed in particular cells. Histone proteins can be altered via post-translational methylation, acetylation, ubiquitination or sumoylation[3] (Gräff and Mansuy 2008). Some, perhaps all, of these changes can be reversible. Environmental factors, including in utero exposure to biomolecules, are now known to have an influence on gene expression via epigenetic modifications.

---

[2]It is important to distinguish between different uses of the term "reductionist". For example, reductionism as a research methodology, in which large problems are broken down into more manageable and testable questions, remains a profoundly valuable tool for the conduct of much science. In contrast, the kind of ontological or causal reductionism, represented here by Crick, is a philosophical position offering (over)simplistic explanations of the functioning of life.

[3]Addition, in order, of a methyl ($-CH_3$ group), acetyl group ($-COCH_3$), a 76 amino acid ubiquitin molecule, or an approximately 100 amino acid Small Ubiqitin-like MOdifier. The details of these processes is irrelevant to the current argument.

3.2.2 Non-coding RNAs (NcRNAs)

Occasionally surprises causes us to fundamentally re-examine areas of biology which we thought were already well understood. One such change has been the revelation regarding the roles played by non-coding RNAs (ncRNAs) in determining whether given protein-coding genes are switched on in particular tissues and/or at particular times.

In a few short years, science has gone from no knowledge about the existence of these small regulatory molecules, to detailed understanding of the roles played by a growing family of ncRNAs including: microRNAs (miRNA), Piwi-interacting RNAs (piRNA), and small-interfering RNAs (siRNA). In our current context, it is sufficient to recognise that these molecules have in common the ability to down-regulate or silence the expression of certain genes. Their actions can be influenced by environmental factors, making them important potential contributors to the G × E interactions (Morris and Mattick 2014).

For example, the body must respond to changes in environmental stress, such as oxygen shortage and nutrient deprivation, and some of the mechanisms for so doing involve microRNAs (Spriggs et al. 2010). There is also growing evidence that environmental factors including cigarette smoke, bisphenol A and exposure to certain metals can lead to alterations in gene expression via pathways involving microRNAs (Hou et al. 2011).

3.2.3 Genes Do Influence Behaviour

Some role for environmental factors is, of course, entirely in keeping with the findings of twin studies, adoption studies and pedigree analysis discussed previously. So too, however, is the expectation that some aspects of behaviour will be influence by genetic criteria. Experiments are starting to identify specific genes implicated in the predisposition to different behavioural abnormalities. These include the genes for brain-derived neurotrophic factor (*BDNF*), neurogenic locus notch homolog protein 4 (*NOTCH4*), neural cell adhesion molecule (*NCAM*) and the serotonin transporter (*5HTT*) (Raine 2008). However, one example, the gene for the monoamine oxidase A (*MAOA*), has been studied in far greater detail than any other, and it is to this that we now turn.

Monoamine oxidase A: Originally found in blood serum, serotonin is a biological molecule with a variety of functions, including a role as both a hormone in the peripheral blood system and as a neurotransmitter within the Central Nervous System (Rang et al. 2007). Later identified chemically as 5-Hydroxytryptamine (5-HT), serotonin is structurally similar to other signalling molecules noradrenalin[4] and dopamine. Abnormally high or low concentrations of these compounds can

[4]Also known as norepinephrine.

have behavioural consequences, including the suggestion that low concentrations of noradrenalin can contribute to depression whereas excessive concentrations are associated with manic behaviour (Rang et al. 2007).

Monoamine oxidase A (MAOA[5]) is an enzyme responsible for inactivation of this class of neurotransmitters, converting them into molecules which, after further processing, are excreted in the urine. Mutations which lead to a low activity or non-functional versions of MAOA will therefore cause neuroactive compounds to be present for an extended period of time.

It has been claimed that in any community, over 50 % of crime will be conducted by fewer than 10 % of the families (e.g. Moffitt 2005). Observation that antisocial behaviour appears to be clustered around certain families added to the long-standing suspicion that genetic factors play a role in conduct of this kind.

This was brought into sharp relief via a landmark study of genetics and metabolism in a Dutch family (Brunner et al. 1993). Several male members of an extended family exhibited impulsive aggressive behaviour accompanied by borderline mental retardation. Analysis of compounds in the men's urine and mapping of their DNA, identified a single letter change (a "point mutation"[6]) within the gene for monoamine oxidase A (*MAOA*). This alteration introduced a premature "stop" signal within the sequence for the protein, leading to production of a shortened and non-functional version. The men with this mutation all had unusually high levels of neurochemicals in their urine and therefore, presumably, within the brain. Female carriers, with one copy of the mutant *MAOA* gene and one normal copy, were unaffected.[7] This correlation of a specific mutation with a behavioural phenotype was hugely significant and has led onto many more nuanced experiments looking into the influence of that mutation as well as a number of other specific changes, some of which are described below.

Studies with transgenic mice: Before returning to consideration of human subjects, in the context of our current discussion, it is worth drawing attention to another study of *MAOA* mutation, this time in mice (Cases et al. 1995). Building upon the work of Brunner and colleagues, an international consortium produced transgenic mice in which the *MAOA* gene had been intentionally inactivated ("knocked out" in the jargon).[8] This was achieved by insertion of a second gene, coding as it happens for interferon beta, within the coding region for monoamine oxidase A, thereby disrupting production of the latter.

[5]By convention the abbreviations for proteins are given in regular text, whereas the abbreviations for genes are put in italics.

[6]Replacement of the usual cytosine (C) by a thymine (T) base at position 936 in the coding sequence.

[7]Females have two X chromosomes; mutation on one copy may be compensated for by a functional copy on the other X chromosome. Because males have one X and one Y chromosome, they only have one copy of genes on the X chromosome and therefore any deficiency will be manifest.

[8]The researchers were intentionally mutating the mouse genome, but via the relatively random mechanism of transposon mutagenesis. The fact that they hit upon the *MAOA* gene was providential and not their initial goal.

Manufacture of these mice allowed for several lines of research. Firstly, it permitted careful monitoring of the concentration of neurochemicals and their breakdown products within different tissues (post mortem), experiments than would not have been possible studying human subjects. Secondly, the researchers were able to study the behavioural effects of this mutation within the mice. Unsurprisingly, they found that male infant mice with an MAOA deficiency had raised levels of noradrenalin, dopamine and particularly serotonin within their brains. On several divergent measures, they also observed that the adult males exhibited more aggressive behaviour than the control mice with the *MAOA* gene intact.

It is important to dwell on the ideological implications of this experiment. Here we see a gene, originally identified via observation of a behavioural trait in humans, for which an identical mutation, deliberately introduced into a lower mammalian model system, leads to similar biochemistry and recognisably similar behaviour. Superficially, this gives credence to the notion that at least some of our conduct is hard-wired; it *appears* to support a deterministic view of behaviour.

Environmental influences on gene expression: All, however, is not quite as straightforward as this observation suggests. A team of researchers, who had already been conducting a prospective longitudinal study[9] on a cohort of over 1000 children in New Zealand, took the opportunity of the identification of the *MAOA* gene to incorporate a genetic dimension into their research (Caspi et al. 2002). Rather than screening for the non-functional version of *MAOA* identified in the Dutch study, the New Zealand team examined the occurrence of a much more common low-activity version of the gene, known as *MAOA-L*. The promoter region of the *MAOA-L* gene, i.e. the section of the gene involved in regulation of the levels of gene expression rather than the sequence of the resultant protein per se, contains 3 copies of a 30 base pair variable-number tandem repeat (VNTR). For whatever reason, this number of copies is associated with low expression levels, whereas an *MAOA-H* version of the gene, which generally has 4 copies of the repeated sequence, has higher expression levels (Buckholtz and Meyer-Lindenberg 2008).

For more than 25 years, the New Zealand investigators had been monitoring instances of maltreatment of the study subjects, and of their participation in anti-social behaviour. When the *MAOA* gene status of the individuals were also into consideration, an interesting pattern emerged. Caspi and colleagues found that simple possession of the *MAOA-L* gene did not correlate with antisocial behaviour. However, men who had this low-expression allele *and* had experienced childhood maltreatment were significantly more likely to be aggressive than males who had the same gene but had not been maltreated. In contrast, if an individual had the *MAOA-H* version of the gene, they did not appear to be as susceptible to the adverse influence of childhood maltreatment (Caspi et al. 2002). The combination of *MAOA-L* and experience of abuse as a child seems to make individuals prone to a

[9]That is to say a long-term investigation of a group, attempting to correlate their experiences with outcomes, without pre-determining the interventions they will encounter.

particular kind of violence where they react disproportionately to some provoking trigger, real or perceived. This ground-breaking observation has been reiterated in subsequent studies (e.g. Kim-Cohen et al. 2006; Frazzetto et al. 2007).

Rather than being a classic case of a gene determining behaviour, it seems therefore that the MAOA story is actually an archetypal example of the importance of genetic *and* environmental factors working together, i.e. a G × E interaction. Roles for epigenetics, and for regulation by ncRNAs, in both the normal and abnormal functioning of the brain are now being postulated. For example, it seems very likely that epigenetic modifications are important in the laying down of long-term memories (Levenson and Sweatt 2005) and miRNAs are reported to have a role in synaptic plasticity and neural development, i.e. in the organisation of the nerves within the brain (Siegel et al. 2011; Qureshi and Mehler 2012).

The best known outworking of epigenetic influences upon brain development involves the curious case of Angelman Syndrome and Prader-Willi Syndrome (Wilkinson et al. 2007). Angelman Syndrome is characterised by the affected individual having mental difficulties but a "sunny" disposition. Prader-Willi Syndrome (PWS) on the other hand leaves sufferers with relatively mild developmental delay, but prone to tantrums and mood swings, and to compulsive eating behaviours. Both conditions most commonly arise from the same deletion on human chromosome 15, which results in several important genes being omitted. As a consequence of differences in the DNA methylation pattern in sperm and egg, children who inherit the copy of chromosome 15 with the deletion from their mother end up with Angelman Syndrome, whereas children who got the mutant copy from their father develop PWS. This process is known as "genetic imprinting".

Of significance to the present book, reports are also starting to emerge of an impact of epigenetic modification on expression of the *MAOA* gene (Shumay et al. 2012). A correlation was seen between the site-specific methylation of the *MAOA* gene promoter and the levels of MAOA activity within the brain.[10]

3.3 Neurophysiology of Self-initiated Action and the Implications for the Understanding of Free Will

Over the past thirty years, experiments investigating the brain activity associated with voluntary decisions to perform a simple movement have been widely reported to undermine the notion of free will. As should be evident, there are certain

[10]Study of CpG methylation was actually conducted using white blood cells, a validated proxy for methylation in the brain. MAOA activity within the brain was monitored via the metabolism of a radioactively-labelled substrate clorgyline, which is only processed by this enzyme. PET was used to record the levels of radioactivity.

complexities in designing an experiment in which you seek to conduct neuroscientific measurements that are linked to an apparently spontaneous action on the part of the research subject (Haggard 2011): How can a scientist with appropriate instrumentation be "on hand" to measure a truly voluntary act? If a volunteer is connected to your electrophysiology apparatus, how can they be expected to demonstrate genuine spontaneity? And how can they indicate the timing of their conscious "will to act" relative to the brain signals that might be recorded, when consciousness is an 'internal' phenomenon shielded from an observer?

The potential limitations regarding transfer of the findings of lab-based research into real-life contexts, the so called ecological validity problem, is one to which we will return in Chap. 5, when we consider the appropriateness of legal application of genetic and neuroscientific data. In the meantime it is necessary to discuss some key experiments looking at the timing and location within the brain for various apparently volitional actions.

3.3.1 Studies Involving Electroencephalography

Although not the first to work on the topic, a classic study was conducted by Benjamin Libet and colleagues (Libet et al. 1982, 1983). Their results appear to show brain activation *before* the participant has made a conscious decision to act. These experiments are worthy of consideration in some detail, because, as noted above, they have been seized upon by advocates of a deterministic worldview to disparage the existence of free will.

Libet was able to construct a system whereby electroencephalographic (EEG) readings were made using electrodes attached to the skull of the participant. His novel contribution was to develop a means whereby the research subject had some autonomy over when to trigger a response (in this case, a short, sharp movement of the right hand), and to report with some accuracy the time at which they were conscious of the intent to do so.

The participant was asked to look at a "clock" (actually a bright dot moving around the edge of an oscilloscope tube) and note the "time" at which they had decided to act. Significantly the research subjects themselves were free to choose how long after the initial set-up they waited to perform the hand movement, so the timing was at least quasi-voluntary.

The self-reported awareness of the time at which participants wanted to move was validated using a second series of recordings in which the subject reported the time at which they were given a small electrical stimuli by the researchers. For comparison, the participants were also required to perform the same hand movement at a pre-set time on the "clock".

A consistent recording in Libet's experiments, subsequently reproduced by numerous other researchers (see Haggard 2008, for review), is the development of an electrical signal, termed the "readiness potential" (RP) or Bereitschaftspotential. The RP starts a second or two before the time of awareness to want to move, which is referred to as "W". W, in turn, precedes the actual movement ("M") by several hundred milliseconds.

3.3.2 Studies Involving Brain Imaging

More recently, researchers have been able to apply emerging techniques such as PET and fMRI (Sect. 3.1) to interrogate the same questions regarding the timing, and more particularly the location within the brain, when an intention to conduct a self-determined movement.

Experiments of this kind have consistently demonstrated readiness potential signals in the supplementary motor area (SMA) of the cerebral cortex (e.g. Jenkins et al. 2000; Cunnington et al. 2002; Lau et al. 2004; Soon et al. 2008; Fried et al. 2011). Attempts to identify the location of activity more finely within the SMA have generated some disagreement, though authors have frequently drawn attention to the importance of the most anterior section, often termed the pre-SMA, in preparation for voluntary action (see Haggard 2008, for review). It is worth noting in passing, however, that there remains some contention regarding the validity of the criteria used to classify sub-regions within the SMA, which may represent a continuum rather than discrete sectors (Nachev et al. 2008).

3.3.3 Investigations at the Neuronal Level

As noted in Sect. 3.1, no individual method for studying the neuroscience of cognition is perfect. EEG-based experiments suffer from a lack of precision regarding the physical location of a signal, whereas fMRI studies are less accurate in regard to the timing of events. Additional information can come from investigations at the level of individual neurons. However, the invasive nature of such work—it involves placing electrodes within the brain—means that studies of that kind have, for the most part, been restricted to animal models (see Rizzolatti et al. 2001, for review). These types of experiments suffer from other complexities. In addition to ethical issues relating to conducting intrusive research on primates (Nuffield 2005), there are evidently practical difficulties in asking the non-human research subject to report on the timings of their intentions.

Rare opportunities to conduct research on human brains do arise, under circumstances in which the tissue has been accessed for another primary goal. For example, Fried et al. (2011) conducted research looking at the activation of single neurons in humans. The experiments involved people who already had electrodes implanted deep into their brains as part of treatment for forms of epilepsy that had proven unresponsive to pharmaceutical interventions.

Their studies found the first measurable neuron level events took place in the supplementary motor areas (SMA) of the cerebral hemispheres. This fits, at least in general terms, with previously published data. The authors are appropriately cautious about the dangers of over-interpretation of their results; they note that the positioning of the electrodes was determined by the clinical needs of the patients and not by their own research priorities. It is therefore feasible that an even earlier signal might have been developed in a different region of the brain, undetected by their study.

Other scientists too have questioned whether it is really likely that an area of the brain responsible for *motor* activity would be the true origin of a *volitional* decision. In a task for which participants were given a choice of two activities (addition or subtraction of two numbers), the researchers were able to tell which choice had been made by examination of the response patterns in the subject's medial prefrontal cortex prior to execution of the decision (Haynes et al. 2007).

As well as activation in the SMA, Soon et al. (2008) used fMRI to demonstrate electrical activity in the frontopolar cortex and the parietal cortex up to 7 s before subjects reported intention to respond (i.e. "W" in the Libet nomenclature). Noting that the fMRI signal can be "sluggish", i.e. slow after the event (relative to EEG), they postulate that these areas may be showing neural activation a full 10 s before W. Timing of this kind would be well beyond any criticisms that self-reporting of intention is sufficiently inaccurate to create a spurious RP effect. It seems therefore that whatever the brain signals preceding conscious intention actually mean, they are nonetheless genuine.

3.3.4 Philosophical and Moral Implications of These Experiments

For present purposes, then, the issue is not so much the details concerning which area(s) of the brain are involved in generating the readiness potential, but rather the implications of the consistent observation of the RP phenomenon. Libet himself has speculated that the delay between the start of RP and the conscious decision to act (W) may represent an opportunity to veto a potential action—not so much "free will" as "free won't" (Libet 1985).[11, 12]

For the majority of determinists, Libet's experiments and subsequent work by others employing a similar research model, show that free will is illusory; an

[11]Despite the interpretations made within the research articles by Libet himself, and by countless others examining the significance of his work, Libet still wishes to retain the potential for free will to be real. In an essay published in 2011, he summarises his position thus: "*My conclusion about free will, one genuinely free in the nondetermined sense, is then that its existence is at least as good, if not a better, scientific option than is its denial by determinist theory. Given the speculative nature of both determinist and nondeterminist theories, why not adopt the view that we do have free will (until some real contradictory evidence may appear, if it ever does)*" (Libet 2011: p 9).

[12]First use of the term "free won't" is generally attributed to Vilayanar Ramachandran but has readily been adopted by Libet and others.

epiphenomenon, *caused by* physical processes in the brain, rather than *causal of* an action. However, some researchers challenge both the traditional interpretation of Libet's data and the significance placed upon them by philosophers.

Roskies notes that the methodology employed in Libert-style experiments skews the outcome towards showing a connection between RP and a movement-based response since the latter is the trigger to the collection of data (Roskies 2011). There may be, she argues, individual RP events that are not associated with movement which are simply not reported due to the way that the experiments have been designed.

Roskies also suspects that the necessity to combine data from large numbers of recordings in order to distinguish genuine waves from background "noise" may lead to significant information being missed. Pockett and Purdy (2011) have gone further in substantiating this criticism by attempting to interpret nearly 400 individual EEG traces. Although the vast majority of traces proved predictably difficult to read "by eye", they found 50 that appeared to have a clear RP and a further 50 that had very low noise but also lacked an RP. Averaging these two populations of data separately confirmed the perception that a sub-population of Libet-style recordings actually have no RP. Given the way the experiment was set up, these readings de facto involved the trigger movement, so it seems that an RP may *not* be a necessary pre-requisite for voluntary action.[13]

Pockett and Purdy also question the received wisdom that an RP is indicative of impending movement. Although given alternative names, e.g. "stimulus preceding negativities" or "contingent negative variations", waveforms have been associated with a range of other circumstances in which the subject is expecting or anticipating something will happen. It may be, they argue, that far from being an initial indication of conscious action, the RP is revealing that the participant is expecting something to happen (albeit that they will have a role within that action).

Roskies also notes that electroencephalography is not a good way to identify the source of a signal. I am not fully persuaded by this argument. The spatial limitations of electromagnetic measurements are well known, but can be overcome by employing different methods. It is noteworthy that her critique makes no mention at all of brain imaging experiments conducted in a Libet-esque manner, despite the fact that these studies, some of which are cited above, have produced copious positional data in support of Libet's basic premise.

Her philosophical concerns are, I think, of greater merit. Firstly, she notes that the set-up of Libet's experiments (and here we could also include the majority of work done in a similar fashion) are actually measuring *when* to move not *whether* to move. This is a relatively trivial question compared with the broader interpretations that have been attributed to the studies. This is not, she continues, an issue that is massively taxing for the conscious will; it is more equivalent to a man deciding to put his trousers on left-leg before right in the morning, rather than to his subsequent

---

[13]The authors are aware of the potential criticism of any experiment with "negative findings" that it may simply not have worked properly.

decision to run a red light in order not to be late for work, which results in a fatal collision at the crossing. "*I can accept that the RP for a finger movement precedes conscious awareness of an intention to move*", she concludes, "*and still deny that this has much to say about whether or not we are free in the cases and contexts in which freedom really does matter*" (Roskies 2011: p 19).

Secondly, Roskies observes that the Libet methodology requires participants to be simultaneously attentive to an external signal (the clock) and an internal signal (intent to move). This may be problematic in terms of the processing of information. Furthermore, what is actually being measured at time W is not *conscious intent*, but *consciousness of conscious intent* which is, potentially, a very different phenomenon.

Philosopher Alfred Mele (Florida State University) is also sceptical about interpretation of the Libet experiments. Amongst other criticism, some of which overlap with Roskies' comments, Mele notes that the data relate to brain activity very close in time to the associated action. In real life, he points out, a particular action may stem from a very long period of deliberation. He concludes that "*on any reasonable conception of free will, the [Libet studies] leave it open both that we sometimes exhibit it and that we never do*" (Mele 2011: p 30).

3.4 Brain Injury: Examples of Altered Behaviour Arising from Changes in Brain Structure

Finally in this chapter, we need to briefly rehearse the potential insights into links between brain structure and behaviour revealed by cases of brain injury. One of the goals of brain imaging research is to identify causal links between brain structure and behaviour. Many investigations, for example, have noted correlation between the volume of brain tissue in given sections of the prefrontal cortex and particular antisocial behaviours (reviewed by Yang and Raine 2009). Studies are increasingly demonstrating the behavioural consequences of changes to an individual's brain. These changes can arise in a number of ways, notably trauma, tumour or atrophy, and each of these will be considered in turn.

3.4.1 Trauma

For more than a century, authors interested in the impact of brain trauma on behaviour have, of necessity, cited the remarkable case of Phineas Gage (e.g. Damasio et al. 1994). Gage was a worker for the Rutland and Burlington Railroad in New England where he was considered reliable and intelligent. He was popular with both colleagues and management. One of his roles within the company was to pack down explosive powder using a metre-long tamping iron. In September 1848, a lapse in concentration saw him attempt to pack-down the explosive before a protective layer of sand had been added. A spark ignited the powder, sending the iron rod through his cheek and out of the top of his skull.

If Gage had died, this would have been the end of the story; just one more work-related fatality in a later 19th Century industrial context. Amazingly, Gage lived through the incident. Whether or not it would be fair to say he 'survived' the accident is a moot point; his persona was radically altered after the event. Although he seemed to have retained his intelligence, speech abilities and physical coordination, he now had no regard for social conventions; he was prone to swearing, irresponsible and untrustworthy. The company had to let him go.

Gage eventually died in the 1860s. No autopsy was conducted at the time but his skull was later recovered and ultimately became the property of Harvard University. This has allowed for periodic reconsideration of the specific damage caused by passage of the tamping rod through Gage's brain (e.g. Damasio et al. 1994; Van Horn et al. 2012). Although there is general consensus that the left frontal lobe was destroyed in the accident, authors differ in their speculations regarding the additional regions affected. In Sect. 3.3 it was noted that the existence of the so-called readiness potential was more important to the current discussion than the exact location of that signal within the brain. Similarly, the precise details of the damage experienced by Gage in the accident are less significant than the uncontroversial observation that severe brain trauma resulted in profound behavioural changes.

3.4.2 Tumour

We have already encountered the case of Charlson (Sect. 2.3), whose physical assault on his son was apparently influenced by a brain tumour. A more recent case is rapidly becoming as notorious as Phineas Gage when considering the psychological impact of physical damage to the brain. During the year 2000, a 40 year old man from Virginia with no previous history of sexual deviancy started to develop a keen interest in pornography, especially images of child (Burns and Swerdlow 2003). In addition to this, he started buying the services of prostitutes and began to make sexual advance towards his prepubescent step-daughter. When the latter reported this to her mother, the man's other activities came to light and he was offered either a residential therapy programme or a prison sentence. He chose the therapeutic alternative, but was expelled from the centre after making sexual advances at female staff members, making a jail term his only option.

On the evening prior to transfer to prison, the man complained of a severe headache and was taken to hospital. Given the circumstances (i.e. imminent incarceration) the doctors suspected the cause of his headache to be psychological rather than physiological. However, the following morning the man also reported balance difficulties and was sent for magnetic resonance imaging (MRI) of his brain. Somewhat unexpectedly, the MRI scan revealed a tumour in the orbitofrontal cortex, an area of the brain known to be involved in regulation of social behaviour.

The tumour was removed, the man underwent a Sexaholics Anonymous course and thereafter returned to the family home. Some while later he developed a persistent headache. He also began to covertly collect pornographic images again. Further MRI demonstrated that the tumour had regrown (February 2002). At the time of his case report (March 2003) the man was no longer prone to deviant sexual behaviour.

3.4.3 Atrophy

With an ever-ageing population, neurodegenerative disorders such as Alzheimer's Disease and Parkinson's Disease are becoming increasingly common. Memory loss is one characteristic associated with these conditions. However, in the context of our present discussion, it is worth noting that behavioural changes can also emerge (McKeith and Cummings 2005). As areas of brain tissue die away, other symptoms can include the emergence of antisocial behaviour, such as sexual disinhibition and increased aggression (Burns et al. 1990). Men carrying the abnormal Huntington's disease gene are disproportionately likely to be involved in criminal behaviour (Jensen et al. 1998).

So, what impact *might* genetic or neurological abnormalities have had on the behaviour of defendants in criminal trials? In the next chapter we will review some of the existing cases in which attempts have been made to introduce neuroscientific and genetic evidence.

References

Alper, J.S. 1998. Genes, free will and criminal responsibility. *Social Science and Medicine* 46: 1599–1611.

Andrews, D.A., and J. Bonta. 2010. *The psychology of criminal conduct*, 5th ed. New Providence, NJ: Matthew Bender & Company.

Balleine, B.W., and S. Killcross. 2006. Parallel incentive processing: an integrated view of amygdala function. *Trends in Neurosciences* 29: 272–279.

Baum, M.L. 2013. The monoamine oxidase A (MAOA) genetic predisposition to impulsive violence: is it relevant to criminal cases? *Neuroethics* 6: 287–306.

Berger, A. 2002. Magnetic resonance imaging. *British Medical Journal* 324: 35.

Bird, A. 2007. Perceptions of epigenetics. *Nature* 447: 396–398.

Brierley, B., P. Shaw, and A.S. David. 2002. The human amydala: a systematic review and meta-analysis of volumetric magnetic resonance imaging. *Brain Research Reviews* 39: 84–105.

Brogdon, B.G., and J.E. Lichtenstein. 1998. Forensic radiology in historical perspective. In *Forensic radiology*, ed. B.G. Brogdon, 18–32. Boca Raton, USA: CRC Press.

Brunner, H.G., M. Nelen, X.O. Breakefield, H.H. Ropers, and B.A. van Oost. 1993. Abnormal behavior associated with a point mutation in the structural gene for monoamine oxidase A. *Science* 262: 578–580.

Buckholtz, J.W., and A. Meyer-Lindenberg. 2008. MAOA and the neurogenetic architecture of human aggression. *Trends in Neurosciences* 31: 120–129.

Bunge, S.A., K.N. Ochsner, J.E. Desmond, G.H. Glover, and J.D.E. Gabrieli. 2001. Prefrontal regions involved in keeping information in and out of mind. *Brain* 124: 2074–2086.

Burns, A., R. Jacoby, and R. Levy. 1990. Psychiatric phenomena in Alzheimer's disease. IV: Disorders of behaviour. *The British Journal of Psychiatry* 157: 86–94.

Burns, J.M., and R.H. Swerdlow. 2003. Right orbitofrontal tumor with pedophilia symptom and constructional apraxia sign. *Archives of Neurology* 60: 437–440.

Cases, O., I. Seif, J. Grimsby, P. Gaspar, K. Chen, S. Pournin, U. Muller, M. Aguet, C. Babinet, J.C. Shih, and E. De Maeyer. 1995. Aggressive behavior and altered amounts of brain serotonin and norepinephrine in mice lacking MAOA. *Science* 268: 1763–1766.

Caspi, A., J. McClay, T.E. Moffitt, J. Mill, J. Martin, I.W. Craig, A. Taylor, and R. Poulton. 2002. Role of genotype in the cycle of violence in maltreated children. *Science* 297: 851–854.

Cunnington, R., C. Windischberger, L. Deecke, and E. Moser. 2002. The preparation and execution of self-initiated and externally-triggered movement: a study of event-related fMRI. *NeuroImage* 15: 373–385.

Damasio, H., T. Grabowski, R. Frank, A.M. Galaburda, and A.R. Damasio. 1994. The return of Phineas Gage: clues about the brain from the skull of a famous patient. *Science* 264: 1102–1105.

Farwell, L., and S.S. Smith. 2001. Using brain MERMER testing to detect knowledge despite efforts to conceal. *Journal of Forensic Sciences* 46: 135–143.

Frazzetto, G., G. Di Lorenzo, V. Carola, L. Proietti, E. Sokolowska, A. Siracusano, C. Gross, and A. Troisi. 2007. Early trauma and increased risk for physical aggression during adulthood: the moderating role of MAOA genotype. *PLoS ONE* 2: e486.

Fried, I., R. Mukamel, and G. Kreiman. 2011. Internally generated preactivation of single neurons in human medial frontal cortex predicts volition. *Neuron* 69: 548–562.

Gilbert, S.J., and P.W. Burgess. 2008. Executive function. In *Current biology*, R110–R114.

Gräff, J., and I.M. Mansuy. 2008. Epigenetic codes in cognition and behaviour. *Behavioural Brain Research* 192: 70–87.

Haggard, P. 2008. Human volition: toward a neuroscience of will. *Nature Reviews Neuroscience* 9: 934–946.

Haggard, P. 2011. Decision time for free will. *Neuron* 69: 404–406.

Haynes, J.-D., K. Sakai, G. Rees, S. Gilbert, C. Frith, and R.E. Passingham. 2007. Reading hidden intentions in the human brain. *Current Biology* 17: 323–328.

Hou, L., D. Wang, and A. Baccarelli. 2011. Environmental chemicals and microRNAs. *Mutation Research* 714: 105–112.

Jenkins, I.H., M. Jahanshani, M. Jueptner, R.E. Passingham, and D.J. Brooks. 2000. Self-initiated versus externally triggered movements. *Brain* 123: 1216–1228.

Jensen, P., K. Fenger, T.G. Bolwig, and S.A. Sørensen. 1998. Crime in Huntington's disease: a study of registered offences among patients, relatives and controls. *Journal of Neurology, Neurosurgery and Psychiatry* 65: 467–471.

Kandel, E.R., J.H. Schwartz, and T.M. Jessell. 2000. *Principles of neural science*, 4th ed. New York, USA: McGraw-Hill Companies.

Kim-Cohen, J., A. Caspi, A. Taylor, B. Williams, R. Newcombe, I.W. Craig, and T.E. Moffitt. 2006. MAOA, maltreatment, and gene-environment interaction predicting children's mental health: new evidence and a meta-analysis. *Molecular Psychiatry* 11: 903–913.

Lau, H.C., R.D. Rogers, P. Haggard, and R.E. Passingham. 2004. Attention to intention. *Science* 303: 1208–1210.

Levenson, J.M., and J.D. Sweatt. 2005. Epigenetic mechanisms in memory formation. *Nature Reviews Neuroscience* 6: 108–118.

Libet, B. 1985. Unconscious cerebral initiative and the role of conscious will in voluntary action. *Behavioral and Brain Sciences* 8: 529–566.

Libet, B. 2011. Do we have free will? In *Conscious will and responsibility*, ed. W. Sinnott-Armstrong, and L. Nadel, 1–10. New York, USA: Oxford University Press.

Libet, B., E.W. Wright Jr, and C.A. Gleason. 1982. Readiness-potentials preceding unrestricted 'spontaneous' vs. Pre-planned voluntary acts. *Electroencephalography and Clinical Neurophysiology* 54: 322–335.

Libet, B., C.A. Gleason, E.W. Wright, and D.K. Pearl. 1983. Time of conscious intent to act in relation to onset of cerebral activity (readiness-potential). *Brain* 106: 623–642.

Mele, A. 2011. Libet on free will: readiness potentials, decisions, and awareness. In *Conscious will and responsibility*, ed. W. Sinnott-Armstrong, and L. Nadel, 23–33. New York, USA: Oxford University Press.

McKeith, I., and J. Cummings. 2005. Behavioural changes and psychological symptoms in dementia disorders. *Lancet Neurology* 4: 735–742.

Miller, E.K., and J.D. Cohen. 2001. An integrative theory of prefrontal cortex function. *Annual Review of Neuroscience* 24: 167–202.

Mobbs, D., H.C. Lau, O.D. Jones, and C.D. Frith. 2007. Law, responsibility, and the brain. *PLoS Biology* 5: e103.

Moffitt, T.E. 2005. The new look of behavioral genetics in developmental psychopathology: gene-environment interplay in antisocial behaviors. *Psychological Bulletin* 131: 533–554.

Morris, K.V., and J.S. Mattick. 2014. The rise of regulatory RNA. *Nature Reviews Genetics* 15: 423–437.

Muehllehner, G., and J.S. Karp. 2006. Positron emission tomography. *Physics in Medicine and Biology* 51: R117–R137.

Nachev, P., C. Kennard, and M. Husain. 2008. Functional role of the supplementary and pre-supplementary motor areas. *Nature Reviews Neuroscience* 9: 856–868.

Noble, D. 2006. *The music of life*. Oxford, UK: Oxford University Press.

Nuffield. 2005. *The ethics of research involving animals*. London, UK: Nuffield Council on Bioethics.

Ollinger, J.M., and J.A. Fessler. 1997. Positron-emission tomography. *IEEE Signal Processing Magazine*, 43–55.

Pockett, S., and S.C. Purdy. 2011. Are voluntary movements initiated preconsciously? The relationships between readiness potentials, urges and decisions. In *Conscious will and responsibility*, ed. W. Sinnott-Armstrong, and L. Nadel, 34–46. New York, USA: Oxford University Press.

Quereshi, I.A., and M.F. Mehler. 2012. Emerging roles of non-coding RNAs in brain evolution, development, plasticity and disease. *Nature Reviews Neuroscience* 13: 528–540.

Raine, A. 2008. From genes to brain to antisocial behavior. *Current Directions in Psychological Science* 17: 323–328.

Rang, H.P., M.M. Dale, J.M. Ritter, and R.J. Flower. 2007. *Rang and Dale's pharmacology*, 6th ed. Edinburgh, UK: Churchill Livingstone.

Rhee, S.H., and I.D. Waldman. 2002. Genetic and environmental influences on antisocial behavior: a meta-analysis of twin and adoption studies. *Psychological Bulletin* 128: 490–529.

Rizzolatti, G., L. Fogassi, and V. Gallese. 2001. Neurophysiological mechanisms underlying the understanding and imitation of action. *Nature Reviews Neuroscience* 2: 661–670.

Roskies, A. 2006. Neuroscientific challenges to free will and responsibility. *Trends in Cognitive Sciences* 10: 419–423.

Roskies, A.L. 2011. Why Libet's studies don't pose a threat to free will. In *Conscious will and responsibility*, ed. W. Sinnott-Armstrong, and L. Nadel, 11–22. New York, USA: Oxford University Press.

Salat, D.H., J.A. Kaye, and J.S. Janowsky. 2002. Greater orbital prefrontal volume selectively predicts worse working memory performance in older adults. *Cerebral Cortex* 12: 494–505.

Shumay, E., J. Logan, Volkow, N.D., and J.S. Fowler. 2012. Evidence that the methylation state of the monoamine oxidase A (MAOA) gene predicts brain activity of MAOA enzyme in healthy men. *Epigenetics*, 7: 1151–1160.

Siegel, G., R. Saba, and G. Schratt. 2011. microRNAs in neurons: manifold regulatory roles at the synapse. *Current Opinion in Genetics and Development* 21: 491–497.

Soon, C.S., M. Brass, H.-J. Heinze, and J.D. Haynes. 2008. Unconscious determinants of free decisions in the human brain. *Nature Neuroscience* 11: 543–545.

Spriggs, K.A., M. Bushell, and A.E. Willis. 2010. Translational regulation of gene expression during conditions of cell stress. *Molecular Cell* 40: 228–237.

Van Horn, J.D., A. Irimia, C.M. Torgerson, M.C. Chambers, R. Kikinis, and A.W. Toga. 2012. Mapping connectivity damage in the case of Phineas Gage. *PLoS ONE* 7: e37454.

Watson, J.D., and F.H.C. Crick. 1953. Molecular structure of nucleic acids: a structure for deoxyribose nucleic acid. *Nature* 171: 737–738.

Wilkinson, L.S., W. Davies, and A.R. Isles. 2007. Genomic imprinting effects on brain development and function. *Nature Reviews Neuroscience* 8: 832–843.

Yang, Y., and A. Raine. 2009. Prefrontal structural and functional brain imaging findings in antisocial, violent, and psychopathic individuals: a meta-analysis. *Psychiatry Research: Neuroimaging* 174: 81–88.

Chapter 4
Use of Genetic and Neuroscientific Evidence in Criminal Cases: A Brief History of "Neurolaw"

Abstract As scientific understanding of behavioural genetics and brain physiology have increased, so too have attempts to utilise this information in criminal cases. Defendants in a growing number of jurisdictions have sought the influence of such factors upon a favourable outcome for their trials or a reduction in the subsequent sentence received. This chapter reviews some of the most prominent international cases in which attempts have been made to introduce genetic or brain imaging evidence into criminal proceedings. The majority have taken place in the USA, which seems to have been quicker to allow consideration of this kind of evidence within the legal system. A willingness to reflect on the merits of evidence of this kind does not necessarily translate into a willingness to let it influence the decisions made. To date, there remain relatively few examples where neuroscientific evidence has had a significant impact on the outcome of the case, but there have been some, and their number is growing.

Keywords Daubert · Death penalty · Fitness to plead · Frye · Genetic predisposition · Ineffective assistance of counsel · Life without parole · *MAOA* · Neurolaw

4.1 The Emergence of Neurolaw in the USA and Beyond

When J. Sherrod Taylor used the term "neurolaw" to describe the expert testimony offered by neurologists to substantiate brain damage in personal injury trials, he can have had little notion that 25 years later Neurolaw would become a field of scholarly endeavour in its own right (Taylor et al. 1991; discussed by Erickson 2010). Although there remain relatively few examples of neuroscientific or behavioural genetic evidence playing a determining role in criminal cases, there is no doubt that such applications are on the increase. The USA appears to have travelled considerably further down this route than most countries. A comprehensive database of American cases, maintained by Professor Nita Farahany at Duke University,

© The Author(s) 2016

C. Willmott, *Biological Determinism, Free Will and Moral Responsibility*, SpringerBriefs in Ethics, DOI 10.1007/978-3-319-30391-8_4

shows that between 2005 and 2012 over 1500 judicial opinions discussed the use of neuroscience by criminal defendants (Bioethics Commission 2015).

This chapter is not the place for an exhaustive review of all case law and attitudes regarding genetic and neuroscientific evidence in disparate countries [readers interested in a recent survey on the latter are directed towards *International neurolaw: a comparative analysis* (Spranger 2012)]. Instead illustrative cases will be drawn from a variety of jurisdictions where relevant. Given the greater use of behaviour-related science in their legal systems, emphasis will inevitably be placed on examples from the USA.

In some senses America is an atypical setting; it is peculiar amongst Western democracies in retaining the death penalty in the majority of states (which, as we shall see below, does have a bearing on some of the applications of brain and behaviour-related evidence). The USA also has a history of tying itself in knots reconciling the admissibility of a range of potential evidence in courts with the Fourth Amendment (the right of the people to be secure in their persons, houses, papers, and effects, against unreasonable searches and seizures) and especially the self-incrimination clause of the Fifth Amendment (the right not to be a witness against oneself in any criminal case).

This latter conundrum is well rehearsed by Farahany (2012) where she reflects on the existing dilemmas regarding the appropriate handling of evidence as diverse as personal papers, unsolicited comments, DNA samples, and blood-alcohol levels, even before the additional quandaries emerging from the introduction of novel neuroscience. Farahany advocates replacing the existing and problematic classification of evidence as either "physical" (admissible despite the Fifth Amendment) or "testimony" (excused under the Fifth Amendment) with a spectrum of "identifying", "automatic", "memorialised" and "uttered" evidence. Since, however, our present focus is on prior cases, this as-yet-untested new model will not be discussed here in any more detail.

4.2 The Place of Scientific Evidence in Criminal Trials

Before we do move on to examine particular cases, it is valuable to lay a certain amount of groundwork regarding the stages of a criminal proceeding at which genetics and neuroevidence might play a part, and to consider more generally the necessary standards regarding scientific evidence and the role of expert witnesses within that process. Reflections on the appropriateness of these applications, and potential future uses of such technologies, will largely be the remit of Chap. 5.

In principal, scientific evidence, including genetic and neuroscientific data, might be presented before, during or after a criminal trial. Evidence could, for example, be presented at a pre-trial competency hearing to ascertain whether a defendant is "fit to plead" at the forthcoming trial. Evidence offered at the trial itself will be used to support or refute the establishment of liability, i.e. guilt for the charges presented. If found guilty, scientific evidence might be introduced at the sentencing phases to help determine the appropriate punishment for the crime. Finally, scientific evidence might again be considered should either side feel there is sufficient reason to launch an appeal against the findings of an earlier trial.

Complex evidence of the kinds described here will, in the majority of cases, be presented by an expert witness. In US federal courts and most state courts it is therefore likely to be subject to Federal Rules of Evidence 702 on the admissibility of expert testimony. These guidelines are often referred to as the "Daubert standard" in light of opinion in a pivotal 1993 Supreme Court case *Daubert v Merrell Dow Pharmaceuticals*,[1] which prompted significant revision to an earlier version of Rule 702. The current (2011) incarnation of the rule states that a witness who is qualified as an expert by knowledge, skill, experience, training, or education may testify in the form of an opinion or otherwise if: (1) the expert's scientific, technical, or other specialised knowledge will help the trier of fact to understand the evidence or to determine a fact in issue; (2) the testimony is based on sufficient facts or data; (3) the testimony is the product of reliable principles and methods; and (4) the expert has reliably applied the principles and methods to the facts of the case (Legal Information Institute, undated). The rule grants a "gatekeeper" role to the trial judge, to try to avoid the presentation of "pseudoscience" and as-yet-unproven science before a jury. This is acknowledged to be a balancing act, since the latter may, at a subsequent time, come to be accepted as the norm within the scientific community.

Some states have never formally adopted any version of Federal Rule 702 or *Daubert*, and instead rely on the much earlier case of *Frye*, or their own state test derived from *Frye*. Concerning the validity of scientific evidence and related testimony, opinion in *Frye* stated that this "*must be sufficiently established to have gained general acceptance in the particular field in which it belongs*".[2]

In the UK, expert witnesses are distinguished by their knowledge or experience of a particular field or discipline beyond that to be expected of a layperson. The most significant difference in function for the court, relative to other witnesses, is that experts are allowed to offer *opinion*, whereas lay witnesses are only allowed to proffer evidence of *fact*.

---

[1]Daubert v Merrell Dow Pharmaceuticals, Inc, 509 US 579 (1993). Jason Daubert and Eric Schuller and their parents sued the pharmaceutical company arguing that the morning sickness drug Bendectin, taken by both mothers during pregnancy, was responsible for the limb deficiencies of their sons. Lower courts had held that epidemiological data offered by the company (showing that mothers who had taken Bendectin were not at a significantly higher risk of having children with birth defects) was more valid than in vitro and animal-model experiments supporting a link. The Supreme Court held that the lower courts had placed too much emphasis on the *Frye* requirement for "general acceptance" and in consequence vacated and remanded this case back to the District Court. In so doing they explicitly stated that the Federal Rules of Evidence superseded *Frye*.

[2]Frye v United States, 293 F 1013 (D.C. Cir. 1923). James Frye, being tried for second degree murdered, attempted to offer in his defence expert testimony interpreting blood pressure measurements as an early form of lie-detector test. The court found the evidence inadmissible.

4.3 Genetics Factors

Although the linking of identified genotypes to behaviour is relatively new (see Sect. 3.2.3) and cases citing a specified gene have, thus far, been extremely rare, appeals to "genetic predisposition" evidenced by family history have become more common, especially in America.

Over a series of three studies, Deborah Denno, Professor of Law at Fordham University, has conducted careful analysis of US criminal cases in which behavioural genetic evidence has been offered (Denno 2006, 2009 and 2011). In the 17-year from 1994 to July 2011, a total of 81 cases citing evidence of this kind were documented in legal databases.[3] Over the course of this period it appears that American courts have become much more willing to accept inclusion of behavioural genetics arguments, especially at the sentencing phase and, even more so, on appeal. Although generally true, this statement does, however, require two important caveats.

Firstly, a distinction needs to be drawn between a willingness to accept genetic evidence as being admissible for *consideration* by the court, and a willingness to be persuaded by such evidence in the *determination* of legal decisions. Greater inclusion of behavioural genetic evidence does not mean that judges will de facto allow it to influence judgements in the manner a defendant might hope.

Secondly, although it does appear that in the recent past judges have been more willing to allow genetic predisposition to be grounds for vacating death sentences, this will not necessarily remain so. Indeed, there is already evidence that ramifications precipitated by *Cullen v Pinholster* (2011)[4] may be reversing this trend.

Scott Pinholster: Scott Pinholster's dealings with the judicial system were long and complicated, and not all of the twists and turns are pertinent to our present discussion. The 'short version' goes as follows. Back in 1984, Pinholster was sentenced to death in a California state court for murder. There followed a protracted series of appeals in which Pinholster argued that there had been inadequate consideration of potential mitigating evidence during the sentencing phase of his initial trial. Eventually a Federal District Court awarded Pinholster the right to a fresh hearing, at which significant additional information regarding both his family history, and evidence of the defendant's brain injuries, were presented. The court granted Pinholster federal habeas[5] relief.

---

[3]LexisNexis and Westlaw databases.

[4]Cullen v Pinholster 131 S. Ct. 1388 (2011).

[5]*Habeas corpus* (from the Latin "you have the body") is a legal procedure whose origins can be traced all the way to the Magna Carta in 1215. In essence it is a safeguard against unfair incarceration, requiring an official such as a prison warden, when so requested, to present their prisoner before a judge for consideration of the legality of their detention. In the context described here, Habeas corpus is used at a federal level to check the legitimacy of a custodial sentence order by a state court.

At the behest of Cullen, Warden of the San Quentin jail, the Supreme Court reconsidered Pinholster's case. They agreed with him that the US *Antiterrorism and Effective Death Penalty Act* of 1996 prevents federal courts from considering habeas petitions based on facts that were available, but were not presented, during consideration at state courts. Accordingly, they reversed the earlier decision.

The reason this is relevant in the current context is not the quality of the genetic and neuroscientific evidence in the *Pinholster* case per se; these were not really discussed. The importance of the case stems from the more general questioning of the legitimacy of federal courts to introduce fresh evidence when reconsidering state court cases. This, Denno (2011) argues, is already having an impact on the capacity of defendants to introduce additional evidence, including family history and behavioural genetics evidence, at later trials.

In acknowledging this second caveat, we have, however, jumped to reflection on present and future restrictions regarding the introduction of genetic and neuroscientific material. Let us turn instead to a potted history of the evolution of the use of evidence of this kind in earlier trials.

Two landmark US cases involving behavioural genetic evidence occurred in the mid-1990s. In the first, Stephen Mobley made an early bid to introduce molecular evidence emerging from contemporary genetics. We will consider *Mobley*[6] in more detail in Sect. 4.3.1 below, alongside other cases where attempts have been made to proffer DNA-based evidence. It is worth noting in passing, however, that in this instance the court found his molecular genetic evidence inadmissible; Mobley was found guilty and ultimately executed.

Susan Smith: The second important case concerned Susan Smith. In 1995, Smith was tried in South Carolina for the murder of her children, who drowned when the car in which they were securely buckled rolled into a lake. Both sides were in agreement that this was no accident. The defence argued that Smith was suicidal because her husband discovered she had been involved sexually with both his boss Cary Findlay, and Findlay's son Tom, and was going to expose her to public disgrace over the affairs. Prosecutors argued that Smith had wanted the children out of the way in order to secure a long-term relationship with Tom Findlay, who was on record as saying he was interested in a future with her, but not one involving her children (Denno 2011).

Central to the defence case was the suggestion that Susan Smith had a genetic predisposition to depression. They presented the court with a genogram, a pedigree diagram in which family members who demonstrated mental instability had been highlighted. Around this, the defence built a fuller picture of Smith's own history of depression and previous suicide attempts. Closing arguments included reference to "her blood line and her genetic inheritance".[7]

This strategy proved partially successful; it did not stop Smith being found guilty of the crime, but she was given a 30-year prison sentence rather than the death

[6]Mobley v State, 455 SE 2d 61 (1995).

[7]State v Smith 1995 WL 789245 (S.C Gen. Sess.).

penalty which was an option available to the court. In subsequent interviews, several jurors indicated that the family history had been prominent in their decision not to recommend execution.

The sparing of Susan Smith represents an unusual success amongst early (in this context, pre-2004) cases (Denno 2006). This may be a consequence of the fact that, at that time, reference to "genetic predisposition" or similar was often made only in passing. In other instances appeals to genetic causes were little more than a proxy for "a family history" of mental illness/alcoholism/depression/etc,[8] without the degree of elaboration exhibited by Smith's defence team. In one case the fact a defendant's step-siblings, brought up in the same abusive environment, did *not* share his inclinations to violent behaviour was offered (unsuccessfully) as evidence of an underlying genetic cause.[9] In another, expert witnesses clashed over whether or not the defendant had physical features that may have been typical of a genetic disorder.[10] In a further case, genetic evidence was considered inadmissible because the court did not believe the social worker advocating genetic explanations for the defendant's alcoholism had sufficient expertise in the field.[11]

Over time, however, there has been a shift to greater willingness to countenance genetic evidence, particularly at the sentencing phase. As Denno (2011: pp. 973–974) notes: "*The question now is not whether courts will accept behavioural genetic factors (they overwhelmingly do), but rather what role those factors will play in particular cases in the context of mitigation evidence*".

In her more recent survey, covering the period June 2007 to July 2011, a total of 33 cases involving behavioural genetics evidence were identified. With one exception, these cases all involved defendants who had been convicted and given a death sentence.[12] In 26 of the cases the defendants' grounds for reconsideration included claims of "ineffective assistance of counsel".[13] In 15 cases, genetics was cited to help substantiate diagnosis of a mitigating condition such as addiction to substance abuse or mental illness. In 10 of the cases, i.e. just under a third, appeal successfully led to the vacating of a death sentence.

---

[8]e.g. State v Ferguson, 20 S.W. 3d 485 (Mo. 2000); State v Hughbanks, 99 Ohio St, 3d 365, 2003-Ohio-4121, 792 N.E. 2d 1081 (2003); Davis v State, M2003-00744-CCA-R3-PC, 2004 WL 253396 (2004).

[9]Cauthern v State, 145 S.W. 3d 571 (Tenn. Crim App. 2004).

[10]Hall v State, 160 S.W. 3d (2004).

[11]People v Armstrong, 700 N.E. 2d 960 (Ill. 1998).

[12]The exception Morris v Malfi No. 10-16485 (9th Cir. Sep 12, 2011) differed from the others on two grounds. Firstly, the appeal was against life imprisonment (without parole) rather than capital punishment. Secondly, this was the only case in which it was argued that the defendant was incompetent at the time of the trial.

[13]The two-part test for the legitimacy of claiming ineffective assistance of counsel was set by the Supreme Court in Strickland v Washington 466 US 668 (1984); (i) Was counsel actually deficient? (ii) Did this deficiency prejudice the outcome of the case?

4.3.1 Molecular Genetics in Criminal Cases

It has been noted many times (e.g. Greely 2011; Walsh and Bolen 2012) that one gene is expressed far more frequently in prisoners than in the general population. The gene in question *SRY* (for Sex-determining Region Y) is located on the Y chromosome, the 'male' chromosome. The reason that it is found expressed in the majority of prisoners is simply a reflection of the fact that most inmates around the world are men. In one sense, therefore, expression of *SRY* correlates with criminal behaviour. However, no-one would seriously advocate that possession of a Y chromosome is sufficient to cause criminality (not least because the majority of men retain the capacity not to indulge in criminal activity).

I mention this here, because attempts to use as mitigation possession of the less common XYY genome, where someone has an *additional* copy of the Y chromosome, have been made. In their review of US case law, Farahany and Coleman (2006) record that on several occasions during the 1960s and 1970s (and as recently as 2003), defendants have attempted to introduce their XYY karyotype as grounds for the insanity that led to their behaviour. Only once, in *People v Farley*,[14] did this argument get as far as consideration before a US jury; they rejected it. In an Australian case of 1968, Lawrence Hannell[15] was acquitted of the murder of a 77 year old landlady. Hannell was found to be XYY, but he was clearly insane according to other M'Naghten criteria (Sect. 2.1.2) and his extra Y chromosome is not considered to have been a significant factor in the decision to acquit him by reason of insanity (Denno 1996). In a French case that same year, the court permitted evidence of an XYY genotype as mitigation when sentencing Daniel Hugon for the murder of a prostitute (Iofrida et al. 2014).

Regarding the use of molecular biology in criminal law, it is important to pick out a number of particular cases where defendants have tried to invoke their *MAOA* genotype. As discussed in Sect. 3.2, monoamine oxidase, the product of the *MAOA* gene, is needed by the body to switch off certain neurotransmitters. In combination with inappropriate childhood stimuli, mutations leading to low level of expression of the gene have been implicated in aggressive behaviour. Each of the cases where the defendant's *MAOA* genotype was cited will be considered in turn.

Stephen Mobley: As noted above, the first attempt to use molecular genetics as mitigating evidence came in the 1994 case of Stephen Mobley.[16] In 1991, Mobley killed the manager of a pizza franchise in Georgia. Following an initial guilty verdict Mobley was sentenced to death. At a Direct Appeal, his defence team requested that their client be allowed to have a genetic test to see if he carried the monoamine oxidase A (*MAOA*) gene mutation which had recently been identified as a factor in the aggressive behaviour of a Dutch family (see Sect. 3.2.3). The

[14]The case [Queens County Sup. Ct, 1827 (1969)] itself is unpublished. However the facts and outcome are described in People v Yukl 372 NYS 2d 31 319-320 (1975).

[15]R v Hannell (unpublished).

[16]Mobley v. State, 455 SE 2d 61 (1995).

judge decided that the courts were not yet ready to consider evidence of this kind and declined their request. Mobley was executed in March 2005 (Eastman and Campbell 2006).

Although Mobley was unsuccessful in his attempt to introduce evidence of a genetic predisposition to aggression, two further cases, both in 2009, saw *MAOA* genotyping used in the courtroom to influence the outcome of trials.

Abdelmalek Bayout: In Italy, Algerian-born Abdelmalek Bayout stood trial for killing another man. Having been verbally abused and then beaten up by a group of young men, Bayout went to a cultural centre and changed clothes, before buying a knife. He set out in pursuit of his attackers, but wrongly identified a man who shared their ethnic background and stabbed him to death (Baum 2013).

At his initial trial, the judge took into account several aspects of mitigation, including the defendant's history of schizophrenia and the fact that he had stopped taking his medication some six months before the crime was committed. The judge deemed these to be grounds for diminished responsibility, but not insanity, and so reduced the sentence to nine years (rather than the usual twelve years). If this was not already sufficiently controversial, it was at the subsequent appeal that genetic evidence was introduced. The defence reported that Bayout possessed the *MAOA-L* allele, encoding a low-activity variant of the enzyme. In light of this additional information, the judge reduced the sentence by a further year, to eight in total (Feresin 2009).

In the current context, there are several important points to note about the Bayout case. Firstly, should failure to take appropriate medication to counter a known medical condition be grounds for reducing a sentence? It might even be argued that this wilful neglect ought to see a sentence *increased*. Secondly, although Bayout had the *MAOA-L* genotype, there was no documented evidence of abuse during childhood (which is considered to be a crucial factor alongside possessions of the relevant allele in the development of aggressive behaviour). Instead, the pertinent environmental influence was taken to be his move from Algeria to Italy at the age of 24 which, it was alleged, led to culture shock and loneliness. Thirdly, there had been a delay of more than one and a half hours between the initial assault and the vengeful murder. This is a longer period between trigger and response than would be usual for impulsive-reactive violence as seen more typically with *MAOA* cases (Baum 2013). Fourthly, in keeping with most of the cases discussed above, the genetic evidence came into play during sentencing, i.e. post-conviction, as a mitigating factor.

Bradley Waldroup: In the other case from 2009, this time in Tennessee, testing for the *MAOA* status of the defendant had an even greater role. During an argument with his estranged wife and her female friend, Waldroup took out his rifle and killed the friend (with whom he believed his wife had been involved sexually). As his wife tried to escape, he shot her in the back and then set about her with other weapons including a knife, a spade and a machete. At this point the police intervened and the wife, though severely injured, was saved.

As with Bayout, *MAOA* evidence was used at Waldroup's trial, but there are several important distinctions between the two cases. Firstly, there was evidence of abuse during Waldroup's childhood, which therefore fits with the pattern revealed by Caspi et al. (2002). Secondly, the fact that Waldroup had the *MAOA-L* allele was raised during the initial, liability trial, i.e. prior to conviction. There was therefore the theoretical possibility that the evidence might have been sufficient to lead to a complete acquittal. As it turned out, the judge reduced the charge (not just the sentence), from first degree murder to voluntary manslaughter. In so doing the sentence was reduce from capital punishment to a maximum of 6 years.[17]

Stefania Albertani: Two years after *Bayout*, the Italian courts were the scene for a second controversial use of genetic (and neurological) evidence. In 2009, Stefania Albertani had pleaded guilty to the murder of her sister. Albertani had been stealing from her family to fuel an addiction to shopping. She killed her sister, who had uncovered evidence of Stefania's financial wrongdoing, and attempted to burn the corpse. Albertani later went on to try and kill her mother who, in turn, was suspicious of Stefania's role in the earlier death (Farisco and Petrini 2012).

Expert witnesses called by the defence (an unusual feature of the case in its own right, since experts are normally appointed by the Court in Italy) conducted both brain imaging and an evaluation of Albertani's *MAOA* status. They reported that there were physical anomalies in the anterior cingulate cortex, and that she had the low-expression allele for *MAOA*. Both of these factors might contribute to aggressiveness and to compulsive behaviour.

The judge in this case was significantly persuaded by the biological evidence, which he described (probably inappropriately) as being less open to subjective interpretation than traditional psychiatric assessment. He therefore did not go with the prosecution's recommendation of life imprisonment and instead sentenced Albertani to 20 years incarceration, with the first three to be served in a mental hospital.

Gary Cossey: Before moving on to look at the use of brain imaging in criminal cases, the case of Gary Cossey[18] is worthy of comment regarding inappropriate arguments based on molecular genetics. The case is unusual both because it was the judge who sought to introduce behavioural genetics, and because the evidence in question was entirely hypothetical. It serves as a warning regarding the scientific literacy of lawyers, a point to which we will return in Chap. 5.

Cossey was found to be in possession of child pornography, a charge to which he pleaded guilty. At the district court, two psychological assessments that Cossey was a low to moderate risk of re-offending were disregarded by the judge who deemed them to be "virtually worthless". Cossey was given a sentence of six and a half years imprisonment followed by lifelong supervision.

It might have been legitimate for the court to query the merit of the psychological evidence in light of the fact Cossey had continued to view child porn, even after being the subject of an initial investigation by the FBI. However the

---

[17]Waldroup in fact received a total sentence of 32 years once other factors, such as kidnapping, were taken into account.

[18]US v Cossey 632 F.3d 82 (2d Cir. 2011).

determination that he ran a substantial risk of recidivism was actually based on spurious genetics.

The court argued that Cossey had an "as yet undiscovered gene", a gene about which more might be known in about 50 years. It was, they contended, "a gene you were born with" and "it's not a gene you can get rid of".

The appeal court found that basing the risk of reoffending on an unsupported theory of genetics was a "plain error"; they vacated the judgement and remanded the case for re sentencing with the stipulation that it was considered by a different judge!

4.4 Use of Brain Imaging in Criminal Cases

Recent years have seen increasing attempts to use data from newer brain imaging methods within the law courts (for more extensive coverage see Jones and Shen 2012; and Rushing 2014). In the context of the current discussion, where the emphasis is biological determinism and moral responsibility, we will omit all civil cases, and criminal cases in which the *plaintiff* has sought to introduce brain imaging data, e.g. as evidence of injury. Only examples in which legal teams have sought to use brain data to provide some explanation for the defendant's behaviour will be considered.[19]

In reviewing the use of neuroimaging in criminal cases it is also important to remind ourselves of the distinction (elaborated in Sect. 3.1.1) between techniques that reveal information about the *structure* of the brain, and those purported to demonstrate brain *function*. A range of approaches including electroencephalography (EEG), Positron Emission Tomography (PET) and functional Magnetic Resonance Imaging (fMRI) offer insight into the operation of the brain. Brain Fingerprinting and the Brain Electrical Oscillations Signature (BEOS) test are variants of EEG.

Our survey of the evolution of the law on insanity, automatism and diminished responsibility (Chap. 2) and on the science of the brain (Chap. 3), have already touched upon cases in which abnormal structures within the brain have been considered to influence the actions of the defendant. In two of these, a brain tumour was deemed to have played a significant role. In the case of the Virginian man with deviant sexual behaviour (Chap. 3), the discovery of a brain tumour was considered to be the root cause of his inappropriate activity and a custodial sentence was reduced to attendance at a therapeutic programme. Earlier we observed that the presence of a tumour led to a verdict of automatism and subsequent acquittal in the case of Charlson (though it is important to be reminded that integral to this outcome was the controversial judgment that the tumour was an "external" not an "internal" factor).

[19]Use by plaintiffs, in an American context, has been exhaustively reviewed by Moriarty (2008).

As noted above, defendants might introduce brain imaging evidence before a trial, in the liability phase or at sentencing (or indeed, all three is they are unsuccessful either in demonstrating that they are incompetent to stand or innocent of the crime). As with our survey of behavioural genetic evidence (Sect. 4.3), review of brain imaging cases here will be illustrative rather than exhaustive. Once again, most examples will be American. We will then conclude this chapter by focusing on two specific issues: the applicability of brain imaging as a tool for lie detection and knowledge of incriminating information (Sect. 4.4.1) and the impact of the study of brain physiology on the culpability of minors (Sect. 4.4.2).

Kenneth Baxter: One surprisingly early example of brain imaging is actually an English case, *Hill v Baxter*[20] dating from 1958. The defendant, Kenneth Baxter, had driven through a Stop sign, whereupon a crash ensued. Baxter told police he had no recollection of the journey leading up to the junction where the accident occurred and believed he was unconscious at the time. The defendant sought, and was granted, permission to introduce electroencephalography (EEG) evidence in support of his claim. In what is now considered an important test case regarding automatism, the appeal court considered the defendant's mental state irrelevant since the issue of importance was his dangerous driving; he had fallen asleep voluntarily and under the law was therefore guilty of recklessness, irrespective of underlying cause. Of relevance here, no comment was made in proceedings regarding the *appropriateness* of the EEG test; the judges it seems were more animated about the apparent slight on their authority given that the consultant neurologist's findings were presented to the court as written evidence rather than in person.

Mohammed Sharif: In another English case,[21] half a century later, it was brain imaging evidence which was ultimately to prove pivotal in the decision that the defendant Mohammed Sharif had not been fit to plead at an earlier trial (Claydon and Catley 2012). At an original 1999 trial, Sharif was found guilty of defrauding the Criminal Injuries Compensation Board by making out that his head injuries following an assault were more serious than was truly the case. Between an initial consideration of his fitness to plead and the first trial both Magnetic Resonance Imaging (MRI) and EEG investigations were undertaken. However, the jury were not persuaded by the medical evidence and found him guilty.

At the appeal, an additional MRI scan from 2000 was considered alongside the earlier brain image. New medical experts felt that the defendant had genuine and progressive cerebral atrophy, but suggested the underlying cause was genetic, the result of too many consanguineous marriages in the defendant's ancestry. Thus although the assault may not have been the cause of the brain damage, the appeal court was persuaded that brain images and expert interpretation showed Sharif had not in fact been fit to plead at the time of the initial trial and quashed the conviction.

[20]Hill v Baxter [1958] 1 All ER 193.

[21]R v Mohammed Sharif [2010] EWCA Crim 1709.

Vincent Gigante: A second case where brain imaging was used in an assessment of competency to stand trial comes from an American context.[22] The case may well have gained particular prominence due to the fact that the defendant, Vincent Gigante, was a Mafia boss. Lawyers for Gigante argued that he was not fit to plead due to his developing dementia. The court ordered that he undergo Positron Emission Tomography (PET, a test of brain function). It would be fair to say that the conduct of this test was less than perfect. In particular, the control group against whom comparison was made, had not been treated with the same psychotropic medicines as the defendant.[23] Nevertheless, the court was not persuaded by the PET evidence and eventually sent Gigante to prison for 12 years.

Andrew Goldstein: Andrew Goldstein was similarly unsuccessful when attempting to use PET data as evidence of insanity.[24] Goldstein had pushed Kendra Webdale under a subway train in an unprovoked assault. All parties agreed Goldstein had carried out the attack and that he was schizophrenic (a diagnosis that pre-dated this offence by some ten years), but was he insane at the time of the incident? A PET scan appeared to demonstrate reduced metabolic activity in his frontal lobe. However the link between this and his insanity defence was not substantiated, and the PET data was excluded from evidence.

Michael Carrizalez: In *California v Carrizalez* (2011),[25] defendant Michael (Miguel) Carrizalez was given a PET scan to investigate any potential relevance of a bullet which had become lodged in his brain following gang-related activity. The PET data was considered at several different stages: pre-trial, during the liability phase and at the sentencing phase of trial.

The imaging data was not deemed sufficient to exonerate Carrizalez, who was found guilty. The PET scan may, however, have had a bearing at sentencing stage, during which the jury failed to achieve the unanimous vote necessary to secure the death penalty. Carrizalez was given a life sentence instead (Rushing 2014).

The fact that the same evidence was introduced at multiple stages of the trial with differing impact in worthy of comment. This case serves as an illustration of the general principal that standards of evidence are interpreted more liberally when deciding on the appropriate punishment for an offence as opposed to the underlying guilt of the defendant (Jones and Shen 2012).[26]

---

[22]United States v Gigante, 982 F. Supp. 140 (E.D.N.Y. 1997).

[23]Problems matching an individual defendant to an appropriate control group is something we will consider again in the next chapter.

[24]People v Goldstein, 14 AD3d 32. Goldstein's conviction was subsequently overturned (People v Goldstein 2005 NY Int. 156), but on the grounds that his rights were violated when a psychiatrist quoted hearsay evidence from a witness who was not made available to the court for cross-examination.

[25]California v Carrizalez (2011), No. VCF 169926C.

[26]It may also demonstrate, and this is entirely speculative, that the psychological burden on jurors of having to choose in favour of the death of another person motivates them to find an alternative if there is opportunity so to do.

Herbert Weinstein: A case in which the defence sought to introduce both structural and functional brain imaging at the liability phase involved 68 year old Herbert Weinstein.[27] Weinstein, with no previous history of violence, was accused of strangling his wife and throwing her from the window of their apartment to give the impression her death was suicide. In the course of various pre-trial testing, MRI (a test of brain *structure*) revealed that Mr Weinstein had an arachnoid cyst. A subsequent PET scan suggested that the region of Weinstein's brain impacted by the cyst was metabolising less glucose than would be anticipated from control experiments with other patients.

The judge noted both that PET was a validated measure of brain activity and that the area affected by Weinstein's cyst is accepted by experts as being responsible for planning and/or self-control. However, he was not persuaded that there is a body of accepted evidence linking this type of brain issue with violent behaviour and thus planned to restrict the interpretation of testimony by the defence witnesses. As it transpired, the defendant changed his plea to guilty of manslaughter at the eleventh hour meaning that the evidence was never actually put before a jury.

Lisa Montgomery: Bobbie Jo Stinnett had met Lisa Montgomery at a dog show in 2004 and they became friends. Most of their subsequent contact was via social media, and during these conversations Montgomery made out that she was at a similar stage of pregnancy to Stinnett.[28] In fact Montgomery was not pregnant at all (she had been sterilised several years earlier), and when they met up (ostensibly so Montgomery could buy a puppy from the other woman) she murdered Stinnett and removed her premature baby by caesarean section, kidnapping the child (a girl). The defence sought to introduce PET evidence to support their case that Montgomery was suffering from pseudocyesis (false pregnancy). The court was persuaded that Montgomery's brain scans showed anomalies in areas related to processing of emotion. However the court decided that there was no proven link between these factors and an insanity plea. Moreover, there was no precedent for this method being used to diagnoses pseudocyesis. They therefore omitted it from the liability phase and found Montgomery guilty. Interestingly, and unusually in capital cases (Rushing 2014) they also excluded this evidence from the sentencing phase, and recommended a death sentence.

Michael Jackson: Michael Jackson[29] was similarly unsuccessful in attempts to employ brain imaging. Jackson admitted shooting police officer Kenneth Wrede but argued brain damage arising from long term use of phencyclidine (PCP) had rendered him incapable of *mens rea*. Expert witnesses for both the prosecution and the defence agreed PET was not a standard method for diagnosis of PCP abuse. This line of evidence was therefore excluded by the court (a decision that was later upheld on appeal).

---

[27]People v Weinstein 591 N.Y.S.2d 715 (N.Y. Sup. Ct. 1992).

[28]United States v Montgomery 635 F.3d 1074 (8th Cir. 2011).

[29]Jackson v Calderon 2000.

Edward Mezvinsky: In *United States v Mezvinsky*,[30] a former Congressman accused of fraud provided a PET scan to support his claim that taking the anti-malarial drug Lariam for several years had caused encephalopathy. The court agreed that the scan showed a decrease in frontal lobe volume but did not agree on the cause of this damage. The fact that both experts, including the witness called by the defence, saw no link between the specific damage and the man's actions did not aid his cause (Moriarty 2008). The court excluded the evidence.

Brian Dugan: In 2009, convicted rapist and murderer Brian Dugan was back in court after DNA evidence confirmed his confession to an additional murder; the brutal killing of a 10-year old girl a quarter-century earlier. During sentencing, Dugan's lawyers sought to introduce fMRI scans in support of their case that he had abnormal brain functioning. Concurring with the lead prosecutor, the judge decided that the images of Dugan's brain might be unduly persuasive to the jury and declined to allow them to be shown in court. He elected instead to allow Kent Kiehl, the defence's expert witness, to describe the evidence, with the aid of cartoon brains and other data representations. Kiehl argued that on a number of different measures, including the fMRI scans, Dugan was clearly psychopathic. The jury were unimpressed by this argument for leniency and unanimously decided in favour of a death sentence[31] (Hughes 2010).

Grady Nelson: When the case of Grady Nelson went to trial in 2010, he appeared to be odds-on to receive the death penalty. Five years earlier, Nelson had brutally murdered his wife Angelina Martinez, stabbing her sixty times and slashing her throat. In fact Nelson left the court facing a life in prison without the possibility of parole. The crucial factor in this outcome was the decision by Judge Hogan-Scola to permit the introduction of quantitative EEG (QEEG) which, she determined, passed both the Frye and Daubert standards for admissibility (Sect. 4.2).

Neuroscientist Robert Thatcher testified that the QEEG tests indicated abnormal activity in the left frontal lobe of Nelson's brain, a region important in the control of behaviour. The damage may have been caused during any of three recorded incidence of brain injury. Experts for the prosecution were far less persuaded by the clarity of the evidence presented, however Judge Hogan-Scola declared that "The methodologies are sound, the techniques are sound, the science is sound" (Gluck 2011).

The jury took only an hour deliberating on the appropriate punishment. Interviewed afterwards about the rationale for their decision, at least two jurors indicated that the QEEG evidence had been pivotal (Miller 2010).

[30]United States v Mezvinksky 206 F Supp 2d 661 (2002).

[31]In 2011, Governor Pat Quinn abolished capital punishment in the State of Illinois so Dugan's sentence was commuted to life imprisonment after all (Zorn, 2011).

4.4.1 Brain Imaging as a Tool for Lie Detection and Knowledge of Case-Related Information

One of the potential applications of neuroimaging techniques which is garnering most interest relates to the possibility of "mind reading". To date courts have been reticent to accept the legal validity of techniques such as Single-photon emission computed tomography (SPECT), fMRI or Brain Fingerprinting for such purposes. Moriarty puts this succinctly, "*they are too new, too uncertain, and too laden with troubling questions to earn easy admission to the courts*" (Moriarty 2008: p 48). Having said that, there are now some reported cases in which "mind reading" has been employed for two different purposes; for cognitive information testing and for lie detection.

Cognitive Information Testing: Cognitive Information Testing (CIT) is a method for determining whether the defendant has any "guilty knowledge", that is to say whether their brain betrays awareness of facts relating to a case which would only be known by the perpetrator of the crime. There are established, and allegedly validated, protocols for probing such knowledge using more traditional polygraph methods (Krapohl et al. 2009). Enthusiasts are looking to utilise the same questioning strategies and prompts, but exploiting the newer tools for capturing the answers.

Terry Harrington: In 2003, Brain Fingerprinting was one of the factors contributing to the quashing of Terry Harrington's conviction for a murder committed 25 years previously. Harrington was shown photographs associated with both the crime scene and with his alibi. Only the latter generated the involuntary P300 brain signals that are said to be indicative of recognition (Sect. 3.1.1). In and of itself this evidence was insufficient to bring about an overturning of the conviction. However news of the findings prompted a witness to admit that he has lied at the original trial and Harrington was freed.

Jimmy Ray Slaughter: A year later, Jimmy Ray Slaughter also turned to Brain Fingerprinting to support his claims that he had been wrongly convicted of the murder of his former girlfriend and their daughter. The brain signals detected by Larry Farwell (principal developer of the technique) concurred with Slaughter's account of events and did not put him at the murder scene (Witchalls 2004). However this data was not deemed sufficient to bring about exoneration and he was executed in 2005.

The appropriateness of this use of Brain Fingerprinting is disputed. Two of Farwell's former colleagues in the initial development of the technique have spoken out against its use in criminal cases. Both Emanuel Donchin and Peter Rosenfeld stand by the validity of the technique for demonstrating the existence of the P300 phenomenon in scientific experiments. They argue, however, that it is insufficiently robust for the legal applications to which it has been put. Critics are particularly concerned that the P300 signal, which most agree is genuine, does not discriminate adequately between *experiential* knowledge, i.e. something that the subject has personally observed, and *content* knowledge, i.e. information they know but not as a result of personal experience (Brown and McCormick 2011).

Aditi Sharma: The first time the related Brain Electrical Oscillation Signature (BEOS) method was used to provide evidence leading to convictions for murder occurred in India in 2008 (Giridharadas 2008). Aditi Sharma, the principal suspect for the poisoning of her former fiancé Udit Bharati, submitted voluntarily to the EEG-based BEOS method which had been developed by Indian neuroscientist Champadi Raman Mukundan. Sharma's brain activity was monitored whilst she was presented with a series of statements relating to the police version of what had happened, interspersed with neutral statements, such as the colour of the sky. In the view of the presiding judge, the brain scans gave clear evidence that Sharma had "experiential knowledge" of the crime. She and her husband Pravin Khandelwal, indicted as an accomplice, were sentenced to life imprisonment. The couple were both later bailed due to concerns about the quality of the evidence against them, though the BEOS method per se was not mentioned in this ruling (Murphy 2009).

Lie detection: As the name implies, lie detection seeks to establish the truthfulness of the defendant, with particular emphasis on whether or not they were involved in the crime under investigation. As with CIT, the difference compared against existing polygraph methods is the intention to directly interrogate the mind of the defendant[32] using neuroimaging rather than monitoring some other physiological changes.

Lorne Semrau: Probably the most important case to date in which a defendant has sought to introduce brain-based lie detection to support their innocence involved psychologist Dr Lorne Semrau.[33] Dr Semrau was the owner of two businesses whose work included providing psychiatric assessments and prescriptions to nursing homes in Mississippi and Tennessee (Shen and Jones 2011). Many of these treatments were billed to Medicare and Medicaid Services. An investigation uncovered systematic overcharging for work conducted by Semrau's companies. The excessive payments totalled $3 million, a fact that was not disputed by the defendant when the numbers were laid bare. At issue was whether or not Semrau has *knowingly* defrauded the government of this money.

At his initial district court hearing, Semrau tried to offer fMRI-based lie detection testing in support for his innocence. The test had been conducted by Dr Stephen Laken, founder and CEO of Cephos (which, alongside NoLieMRI, is one of the two main companies advocating brain imaging as a tool for examining truthfulness). A series of Specific Incident Questions were developed and put to Semrau whilst he lay in the scanner. Some of the questions related specifically to the case, others were "neutral" questions against which to compare the brain images.

---

[32]In principle the same tools could be used to examine the truthfulness of key witnesses, or indeed to root out potential prejudice being concealed by judges or jurors. These latter applications have not yet reached the courts.

[33]United States v Semrau 693 F.3d 510 (6th Cir. 2012).

As originally intended, Semrau completed two series of questions. Somewhat unhelpfully for his case, the first set of these initial tests showed Semrau was telling the truth, but the second suggested he was lying. Controversially, Cephos agreed to redo the second test on grounds that Semrau may have been fatigued due to taking both tests back to back. The third test now endorsed the first, that Semrau was telling the truth. The defence therefore sought to introduce Laken as an expert witness in the liability phase of the trial to comment on the findings of the assessments (whilst themselves noting that Laken was not directly discussing Semrau's mental state at the time of the financial offences, which had occurred five and more years before the fMRI was conducted).

Judge Tu Pham was called upon to adjudicate over whether the fMRI evidence was admissible. He considered it under Federal Rules of Evidence 702 (the *Daubert* test, see Sect. 4.2) and determined that this application of fMRI technology was unproven in the "real world", in contrast to laboratory conditions.[34] The district court therefore declined to allow this data to be included in Dr Semrau's defence.

On appeal, the Supreme Court endorsed the reticence demonstrated by the lower court. Crucially, therefore, *Semrau* enshrines the ruling that fMRI evidence is currently inadmissible as an indicator of truthfulness in criminal trials in the USA.

Theodoros Krallis: A second case where brain imaging has been used as a tool for lie detection comes from Greece. It is reported that defendant Theodoros Krallis, accused of being an accomplice in the abduction and murder of an entrepreneur, was given permission by a court in Athens to undergo Event-Related Potential (ERP) brain analysis to support his claim of innocence (Vidalis and Gkotsi 2012; Gina Gkotsi, personal communication). There was initially some reticence to permit this application due to Greek legislation protecting defendants from lie-detector evidence used in conjunction with torture. However the defence team argued that Krallis was entering voluntarily into this assessment. He was eventually acquitted, though no reference regarding the lie-detector test was made in the decision.

It is worth noting in passing that brain imaging techniques are unlikely to be employed in UK courts in the foreseeable future, given the existing ban on more established polygraph testing in criminal cases[35] (Archbold 2014).

[34]They additionally decided that the evidence was not admissible since the prosecution had not been given prior notification of this line of defence. They thus had been offered no opportunity to scrutinise the conduct of the test which had been a risk-free strategy by Semrau (since he would not have presented the data if the results had not been in his favour, and the Government lawyers would have been none the wiser).

[35]The use of polygraphs in the ongoing monitoring of convicted sex offenders is currently being trialled in the UK, but there remains no intention to expand this use into courtrooms.

4.4.2 Brain Physiology and the Culpability of Minors

A series of recent Supreme Court cases in the USA have seen significant movement away from the applicability of the death penalty in a variety of contexts. The principal driver for these changes has been concern about violations of defendants' rights, under the Eighth Amendment, not to receive excessive, cruel or unusual punishment. The landmark cases have included *Atkins v Virginia*,[36] in which a majority decision ruled it unlawful to impose the death penalty on intellectually disabled individuals, and *Kennedy v Louisiana*,[37] which outlawed capital punishment for crimes that did not lead to the death of the victim.

Three further cases—*Roper v Simmons*,[38] *Graham v Florida*[39] and *Miller v Alabama*[40]—concerned the treatment of individuals who were not adults at the time of their offence. These will be discussed here in some detail, due to the significant role played by brain science in informing these decisions. Before examining the reasoning articulated by the Supreme Court in these cases, which flow nicely one from another, the initial offences for which the defendants were found guilty will be outlined.

Christopher Simmons: In September 1993, Christopher Simmons and a friend kidnapped Shirley Crook from her home in the middle of the night, bound her with duct tape and electrical cable and threw her into the Meramec River, Missouri where she drowned. There is no question that Simmons was guilty of premeditated murder; before the event he had spoken openly about his fantasy of throwing someone off a bridge. The day after the killing, Simmons was bragging about his exploits and was almost immediately arrested. Within hours he was offering to do a videoed re-enactment for the police.

Terrance Graham: In July 2003, Terrance Graham and three friends attempted to rob a restaurant where one of them worked. The burglary was unsuccessful; they left empty-handed, but not before the restaurant manager had been hit on the head with a metal bar, sustaining injuries that required medical treatment.

Although only 16 at the time, Graham was tried in an adult court on two charges, the more serious of which was armed burglary with assault, carrying a maximum sentence of life imprisonment without the option of parole ("LWOP"). Graham made a plea bargain; for pleading guilty and having declared on oath that he would not take part in any further criminal behaviour, he received a suspended sentence, which included a one year jail-term (some of which he had served prior to the trial).

---

[36]Atkins v Virginia 536 US 304 (2002).

[37]Kennedy v Louisiana 554 US 407 (2008). An exception was made for crimes against the state, such as treason or spying, for which the death penalty could still be invoked.

[38]Roper v Simmons 543 US 551 (2005).

[39]Graham v Florida 560 US 48 (2010).

[40]Miller v Alabama 567 US (2012). Despite the naming of this ruling, the unrelated circumstances of a second appellant Kuntrell Jackson was also considered at the same time.

Notwithstanding his earlier promise, Graham became embroiled in a series of house robberies where he and his partners were armed. In December 2004 he was arrested after dropping an injured accomplice at the hospital. He ignored the call from a police officer to stop, but crashed as he attempted to get away and was apprehended. Returning to court on new charges, he denied involvement in the burglaries, but admitted fleeing from a police officer. This latter admission was sufficient to constitute violation of the probation conditions relating to his earlier trial and he was duly issued with the maximum sentence related to the original charges, namely LWOP. Graham appealed to the district court in Florida, but they found that the punishment was not disproportionate given the seriousness of the charges.

Evan Miller: Evan Miller had a dysfunctional upbringing, involving periods of time in foster care to offer respite from his mother's problems with alcohol and drugs, and abuse at the hands of his stepfather. On the relevant night in 2003, aged 14, he was however living at home in a trailer park. He and a friend had gone to the neighbouring caravan of an older man Cole Cannon, where they had all smoked cannabis and taken part in drinking games. When Cannon fell asleep, Miller stole his wallet, sharing the money with his friend. However as they attempted to return the empty wallet into Cannon's pocket, he woke up and started to throttle Miller. The friend grabbed a baseball bat and hit Cannon with it, causing him to lose his grip on Miller. The latter then took the bat and struck the victim multiple times. Before dealing a final blow, Miller is reported to have placed a sheet over Cannon's head and declared "I am God, I've come to take your life".[41] The boys left, but returned later and set fire to the caravan to try and conceal their crime. Cannon died from a combination of head injuries and smoke inhalation.

Although a juvenile at the time of the crime, Miller's case was move to an adult court by the District Attorney who considered him mature for his age, and due to previous criminal activities (albeit relatively minor). Under these circumstances, a guilty verdict for murder led to a minimum sentence of LWOP.

Kuntrell Jackson: Although not named in the headline of *Miller v Alabama*, the same Supreme Court hearing considered the circumstances of Kuntrell Jackson. Jackson's case was unrelated to that of Miller, in the sense that it took place four years earlier, in Arkansas. However Jackson's initial case had similarities to *Miller* since he too was aged 14 at the time of the offence, but was tried in an adult court and received a sentence of LWOP. It was therefore appropriate for the Supreme Court to deliberate on them at the same time.

In 1999, Jackson and two older boys had set out to steal from a video store. On their way there, Jackson discovered that one of the others had a shotgun concealed in his coat. It is undisputed that Jackson initially stayed out of the store. It is also agreed that it was not him that fired the gun in, what turned out to be, the fatal injuring of the store clerk, Laurie Troup. He had, however, gone into the building by the time the shooting occurred and had said either "we ain't playin" (taken as a

[41]Quoted in Miller Opinion of the Court, p5.

threat) or possibly "I thought you all was playin" (which would have indicated that the situation had escalated beyond what he had expected).[42] Whichever the intent had been, Jackson was nonetheless tried as an accessory to murder. At the discretion of the Arkansas prosecutor, the case was moved from the juvenile to the adult court. When a guilty verdict duly followed, the minimum sentence available was life imprisonment without parole.

Evolution of laws relating to minors: Consideration of these cases by the Supreme Court has led to reversal of a trend in the 1990s during which sentencing for juveniles had become harsher. It has been argued elsewhere that the ratcheting up of punishment in that period was prompted by "moral panic" at the lawless behaviour of teenage "super-predators" (Scott 2013). Importantly, in the current context, insights from recent brain research have been integral to the arguments which have led to this more sympathetic treatment of minors in the judicial system.

There is no question that Christopher Simmons had committed a heinous crime, but was the death penalty appropriate? Not according to the Supreme Court in *Roper v Simmons* where it was decided, by a narrow 5–4 majority, to ban capital punishment for defendants who were younger than 18 at the time of their offence.

Roper was not the first time the Supreme Court had considered the applicability of the death penalty for juvenile offenders. In *Thompson v Oklahoma*[43] (1988) they had overturned the death sentence for a 15 year old on the grounds that it had breached his Eighth Amendment rights. However a year later, in *Stanford v Kentucky*[44] the Court had voted 4–5 in favour of retaining the possibility of executing defendants over 16.

Although the formal Opinion of the Court in *Roper* did not directly refer to any neurological evidence in their justification for changing the legislation, the first of their three key differences between adults and those under 18 noted "*as any parent knows and as the scientific and sociological studies respondent and his amici*[45] *cite tend to confirm, "[a] lack of maturity and an underdeveloped sense of responsibility are found in youth more often than in adults and are more understandable among the young. These qualities often result in impetuous and ill-considered actions and decisions.*"'"[46]

The *amici curiae* in that case included reference to MRI-based work by Paus and colleagues which had shown that myelination of nerve fibres within the brain was a slower process than previously recognised; white matter density was increasing in an age-related fashion up to and including 17 years of age (Paus et al. 1999). From

[42]There are echoes here of the infamous English case of Derek Bentley who was hanged in 1953 for his part in the murder of a police officer. It was undisputed that the fatal shot in that case had been fired by his accomplice Christopher Craig, but the phrase "Let him have it, Chris" was interpreted by the court to be an instruction to shoot rather than, as the defence contended, a suggestion that the weapon should have been surrendered to the policeman.

[43]Thompson v Oklahoma 487 US 815 (1988).

[44]Stanford v Kentucky 492 US 361 (1989).

[45]*Amici curiae* are unsolicited offers of information given to the court by interested parties.

[46]Roper Opinion of the Court, p15. Internal quote comes from Johnson v. Texas, 509 U.S. 350.

these imaging studies it was inferred that the brains of adolescents are still developing and therefore they were not sufficiently mature to be faced with capital punishment.

The Opinion (delivered on behalf of the Court by Justice Kennedy) continued "*If trained psychiatrists with the advantage of clinical testing and observation refrain, despite diagnostic expertise, from assessing any juvenile under 18 as having antisocial personality disorder, we conclude that States should refrain from asking jurors to issue a far graver condemnation—that a juvenile offender merits the death penalty.*" (pp 19–20).

So in the evolution of legislation regarding minors, *Roper* set 18 as the minimum age at which being found guilty of murder could lead to the death penalty.[47] Following on from this, *Graham v Florida* established an outright ban on sentences of life without parole for non-homicide crimes. This decision was based on much the same reasoning as had been foundational in *Roper*, but was put more overtly in *Graham*. Majority Opinion in the latter, given again by Justice Kennedy, stated "*As petitioner's* amici *point out, developments in psychology and brain science continue to show fundamental differences between juvenile and adult minds. For example, parts of the brain involved in behavior control continue to mature through late adolescence*".[48, 49]

Most recently, *Miller v Alabama* ruled out *mandatory* LWOP sentences for defendant aged under 18 at the time of the offence. This case is distinct from both *Roper* and *Graham* since it did not lead to an outright ban; it is still possible for courts to impose LWOP for murder if they feel the circumstances warrant such a sentence, but this cannot be the only available outcome. However, *Miller* is very much on the same trajectory as *Roper* and *Graham* because the reasoning from both cases was quoted in the majority Opinion (this time proffered by Justice Kagan). A footnote in the Opinion (p5) notes that "*The evidence presented to us in these cases indicates that the science and social science supporting* Roper*'s and* Graham*'s conclusions have become even stronger*" going on to specifically cite *Amicus Curiae* from the American Psychological *Association* "*It is increasingly clear that adolescent brains are not yet fully mature in regions and systems related to higher-order executive functions such as impulse control, planning ahead, and risk avoidance*".

In essence, therefore, the Supreme Court appears to have taken the view that brain research is endorsing a belief that the immaturity of their brains makes young offenders both less culpable for their behaviour than adults, and offers greater promise for rehabilitation and an end to their criminal activities. Although the UK and other European countries do not presently countenance the death penalty for

---

[47]As noted above, the Supreme Court had previously set 16 as the minimum, Thompson v Oklahoma 487 US 815 (1988).

[48]Graham Opinion of the Court, p17.

[49]Appeal to neuroscience was even more overt in Sullivan v Florida. However the Supreme Court ultimately decided that Sullivan was improvidently passed to them for consideration and declined to make a judgement. The details have therefore been omitted from the present discourse.

anyone, let alone minors, it is just worth noting that the average age of criminal responsibility in Europe is 14 years, and in England it is 10. Some have argued that the same brain experiments that had contributed to changes in the law regarding juvenile criminals in America ought also to influence of the treatment of children in other legal systems. Sarah-Jayne Blakemore, Professor of Cognitive Neuroscience at University College London and a leading world authority on the adolescent brain, has called for the active input of neuroscientific findings into the legislation regarding appropriate treatment of juveniles. She notes "*The plentiful data that consistently paint a picture of the adolescent brain as relatively immature might speak against the relatively young age of criminal responsibility and harsh sentences for adolescents.*" (Blakemore 2012: p. 404).

What are we to make of the other genetic and neuroscientific evidence outlined in Chap. 3? Does it really warrant the existing legal applications surveyed in this chapter? How should it influence further developments? These are questions to which we turn in our final chapter.

References

Archbold, J.F. 2014. *Archbold criminal pleading, evidence and practice 2014*. In ed. P.J. Richardson, 8–291, 10–58. London: Sweet and Maxwell Ltd.

Baum, M.L. 2013. The monoamine oxidase A (MAOA) genetic predisposition to impulsive violence: is it relevant to criminal cases? *Neuroethics* 6: 287–306.

Bioethics Commission. 2015. Gray *matters (volume 2): topics at the intersection of neuroscience, ethics, and society*. Available online at http://bioethics.gov/sites/default/files/GrayMatter_V2_508.pdf. Last accessed 8 Aug 2015. Washington DC, USA: Presidential Commission for the Study of Bioethical Issues.

Blakemore, S.-J. 2012. Imaging brain development: the adolescent brain. *Neuroimage* 61: 397–406.

Brown, T.R., and J.B. McCormick. 2011. New directions in neuroscience policy. In *Oxford handbook of neuroethics*, ed. J. Illes and B.J. Sahakian, 675–700. Oxford: Oxford University Press.

Caspi, A., J. McClay, T.E. Moffitt, J. Mill, J. Martin, I.W. Craig, A. Taylor, and R. Poulton. 2002. Role of genotype in the cycle of violence in maltreated children. *Science* 297: 851–854.

Claydon, L., and P. Catley. 2012. Neuroscientific evidence in the english courts. In *International law: a comparative analysis*, ed. T.M. Spranger, 305–328. Heidelberg: Springer.

Denno, D.W. 1996. Legal implications of genetics and crime research. In *Genetics of criminal and antisocial behaviour*. Chichester: Wiley.

Denno, D.W. 2006. Revisiting the legal link between genetics and crime. *Law and Contemporary Problems* 68: 209–258.

Denno, D.W. 2009. Behavioral genetics evidence in criminal cases: 1994–2007. *The impact of behavioral sciences on criminal law*, 317–354. New York: Oxford University Press.

Denno, D.W. 2011. Courts' increasing consideration of behavioral genetics evidence in criminal cases: results of a longitudinal study. *Michigan State Law Review* 2011: 967–1047.

Eastman, N., and C. Campbell. 2006. Neuroscience and legal determination of criminal responsibility. *Nature Reviews Neuroscience* 7: 311–318.

Erickson, S.K. 2010. Blaming the brain. *Minnesota Journal of Law, Science and Technology* 11: 27–77.

Farahany, N.A. 2012. Incriminating thoughts. *Stanford Law Review* 64: 351–408.

Farahany, N.A., and J.E. Coleman. 2006. Genetics and responsibility: to know the criminal from the crime. *Law and Contemporary Problems* 69: 115–164.

Farisco, M., and C. Petrini. 2012. The impact of neuroscience and genetics on the law: a recent Italian case. *Neuroethics* 5: 317–319.

Feresin, E. 2009. Lighter sentence for murderer with 'bad genes'. *Nature News*. doi:10.1038/news.2009.1050. Last accessed 14 Sept 2012.

Giridharadas, A. 2008. India's use of brain scans in courts dismays critics. *New York Times*, 15 Sept 2008. Available online at http://www.nytimes.com/2008/09/15/world/asia/15iht-15brainscan.16148673.html. Last accessed 23 Sept 2012.

Gluck, G. 2011. QEEG accepted in death penalty trial in Florida v. Nelson. *Biofeedback* 39: 74–77.

Greely, H.T. 2011. Neuroscience and criminal responsibility: proving 'can't help himself' as a narrow bar to criminal liability. In *Law and neuroscience*. New York: Oxford University Press.

Hughes, V. 2010. Head case. *Nature* 464: 340–342.

Iofrida, C., S. Palumbo, and S. Pellegrini. 2014. Molecular genetics and antisocial behaviour: where do we stand? *Experimental Biology and Medicine* 239: 1514–1523.

Jones, O.D., and F.X. Shen. 2012. Law and neuroscience in the United States. In *International law: a comparative analysis*, ed. T.M. Spranger, 349–380. Heidelberg: Springer.

Krapohl, D.J., J.B. McCloughlan, and S.M. Senter. 2009. How to use the concealed information test. *Polygraph* 38: 34–49.

Legal Information Institute, undated. Rule 702. Testimony by Expert Witnesses. Available online at https://www.law.cornell.edu/rules/fre/rule_702. Last accessed 30 April 2015.

Miller, G. 2010. Brain exam may have swayed jury in sentencing convicted murderer. *Science insider*. Available online at http://news.sciencemag.org/technology/2010/12/brain-exam-may-have-swayed-jury-sentencing-convicted-murderer.

Moriarty, J.C. 2008. Flickering admissibility: neuroimaging evidence in U.S. courts. *Behavioural Sciences and the Law* 26: 29–49.

Murphy, E.R. 2009. Update on Indian BEOS case: accused released on bail. *Law and Bioscience Blog*. Available online at https://law.stanford.edu/2009/04/02/update-on-indian-beos-case-accused-released-on-bail. Last accessed 24th Feb 2016.

Paus, T., A. Zijdenbos, K. Worsley, D.L. Collins, J. Blumenthal, J.N. Giedd, J.L. Rapoport, and A. C. Evans. 1999. Structural maturation of neural; pathways in children and adolescents: in vivo study. *Science* 283: 1908–1911.

Rushing, S.E. 2014. The admissibility of brain scans in criminal trials. *Court Review* 50: 62–69.

Scott, E.S. 2013. Miller v Alabama and the (past and) future of Juvenile crime regulation. *Law and Inequality* 31: 535–558.

Shen, F.X., and O.D. Jones. 2011. Brain scans as evidence: truths, proofs, lies and lessons. *Mercer Law Review* 62: 861–883.

Spranger, T.M. 2012. *International law: a comparative analysis*. Heidelberg: Springer.

Taylor, J.S., J.A. Harp, and T. Elliott. 1991. Neuropsychologists and neurolawyers. *Neuropsychology* 5: 293–305.

Vidalis, T., and G.-M. Gkotsi. 2012. Neurolaw in Greece: an overview. In *International law: a comparative analysis*, ed. T.M. Spranger, 179–196. Heidelberg: Springer.

Walsh, A., and J.D. Bolen. 2012. *The neurobiology of criminal behavior: gene-brain-culture interaction*. Farnham: Ashgate Publishing.

Witchalls, C. 2004. Murder in mind. *The Guardian*, 25 March 2004. Available online at http://www.guardian.co.uk/science/2004/mar/25/crime.uknews/print. Last accessed 23 Sept 2012.

Chapter 5
Are We Ready for an Expanded Use of Neuroscientific Evidence in the Courtroom?

Abstract Evidence in the form of behavioural genetics and brain imaging has started to reach the courtroom. In this concluding chapter, the underlying validity of these methods will be examined. After review it will be determined that electroencephalography, positron emission tomography and functional magnetic resonance imaging are all appropriate techniques for examining the working brain. Despite being scientifically valid, however, it does not follow automatically that the uses for which such evidence has been offered in criminal cases were necessarily justified. Based on current experience, judiciaries would be wise to wait for more robust validation of neurobiological evidence before expansion of its use. This does not mean, however, that data derived via these techniques will not be integral to criminal proceedings in the future. Before taking that step, more closely matched reference populations need to be established, and the interaction of environmental stimuli alongside genetics needs to be better understood.

Keywords Brain overclaim syndrome · Christmas tree effect · Determinism · Ecological validity · fMRI · G2i · Moral responsibility · Reverse inference

Developments in understanding of the genetics and neuroscience underlying behaviour have been amongst the most fruitful areas of recent scientific endeavour. The potential philosophical and legal implications of these discoveries are self-evident; if science were to show that our behaviour is entirely determined by biological factors this would undermine traditional notions of free will and, in turn, the concept of moral responsibility and culpability that underpins the legal system. As Roskies notes, "*If moral responsibility is found to be incoherent, then our social, moral and legal systems will be eviscerated and the result will be chaos.*" (Roskies 2006: p. 421).

Philosophically, it is only hard determinism[1] (Sect. 1.3) which would rule out any notion of moral responsibility, since compatibilism retains scope for personal accountability. Evidence that severed the connection between someone's actions,

[1]Or, conversely, a belief that life is entirely driven by random forces, a view that is untenable for the vast majority of people.

© The Author(s) 2016
C. Willmott, *Biological Determinism, Free Will and Moral Responsibility*, SpringerBriefs in Ethics, DOI 10.1007/978-3-319-30391-8_5

their moral responsibility and/or the appropriate punishment would have seismic effects on the organisation of democratic societies. Bear in mind, as well, that it might not even be necessary for the science to *actually* substantiate this attack on responsibility in order for the chaos described by Roskies to ensue. If there is sufficient popular *perception* that our behaviour is in some way hard-wired, and therefore outwith our conscious control, then trust in the legal process might be undermined regardless. The hypothesis that developments in neuroscience challenge traditional views regarding the existence of free will has certainly reached some quarters of the popular consciousness (e.g. Economist 2006; Chivers 2010; Gutting 2011; Stafford 2015).

Drawing together our reflections on the philosophical, legal and scientific dimensions of moral responsibility, we are led towards two bipartite questions. Firstly, is the genetic and brain analysis data regarding behaviour scientifically robust and, if it is, does the science support a deterministic worldview? Secondly, should the science be used in the courtroom and, if so, in what ways? Each of these issues will be considered in turn.

5.1 Is the Genetic and Brain Imaging Evidence on Behaviour Scientifically Robust and, if so, Does the Science Support a Deterministic Worldview?

If it was established that the science of behaviour could be accurately described via the study of neuroscience and genetics, it would not necessarily follow that this evidence ought immediately to play a role in the legal system. However, if the science was shown to be fundamentally flawed it would argue strongly against its use in the context of criminal justice.

The kinds of research being conducted to investigate the biological basis of behaviour have been discussed at length in Chap. 3. Here we examine specifically issues regarding the validity and reliability of this data considering, in turn, the brain imaging and the genetic approaches.

5.1.1 Neuroscience

Over the past century a variety of approaches to brain imaging have been developed (Table 3.1). In terms of potential applicability in the criminal court, it is the scientific validity of Electroencephalography (EEG), Positron Emission Tomography (PET) and functional Magnetic Resonance Imaging (fMRI) which need to be examined most closely.

EEG: After more than 80 years of use and refinement, the value of EEG as a method for examining cognitive processes is now well established (Lopes da Silva

2013). Recent developments are even bolstering the spatial as well as the temporal dimensions of EEG (Michel and Murray 2012). It remains, of course, possible to abuse a scientifically-robust technique by misapplication of the data generated and some controversy persists regarding the legal application of EEG-generated evidence as an assessment of incriminating memories (Sect. 4.4.1). The method itself is, nevertheless, considered reliable.

PET: As with EEG, PET is also considered to be a scientifically-validated method, with the highest sensitivity of any technique for measuring specific ligand-receptor interactions in situ (Zimmer and Luxen 2012). The use of glucose labelled with radioactive fluorine to study brain metabolism has been extensively researched and the underlying science is well understood.

The involvement of radioactivity is both a strength and a weakness of PET. The use of purposely-labelled molecules gives more straightforward signals than other methods such as fMRI (as we will see in a moment). However, the health risks associated with ionising radiation also limits their usage.

fMRI: Of these three techniques, therefore, it is the scientific validity of fMRI which requires the closest inspection. We live in an era where neuroscientific spin is being applied to an increasing number of fields. Aside from Neurolaw, there is growing interest in areas such as Neuromarketing (e.g. Ariely and Berns 2010), Neuropolitics (e.g. Henry and Plemmons 2012) and Neuroeconomics (e.g. Glimcher and Rustichini 2004).

Increased use of fMRI is at the core of this spread into different disciplines, which Geoffrey Aguirre, Neurologist at the Perelman School of Medicine has termed "imaging colonisation" (Aguirre 2014). As is often the case when an innovation carefully nurtured in one community is adopted by another, without the requisite history, there are concerns that new adopters may use the technology inappropriately and overstate the results. In keeping with the nomenclature, this has led to coining of the terms "Neuroskepticism" (e.g. Marks 2010; Rachul and Zarzeczny 2012) and even "Neurobollocks" (e.g. Poole 2012).

In a thorough and balanced review, University of Pennsylvania neuroscientist Farah (2014) has examined whether or not the fundamental criticisms of the science of fMRI (as opposed to their application in criminal cases) are warranted. A first area of concern emerges from the indirect manner in which the image data is generated. As noted in Sect. 3.1.1, fMRI actually reports on the magnetic properties of haemoglobin, which is influenced by the levels of oxygen within the blood, as a proxy for neuronal activity. Processing of information by the brain needs energy and the flow of blood bringing oxygen to the site serves a second-hand report of the action. Clearly this is not ideal, we would like to be able to "see" the nerve cells directly rather than measuring blood-oxygen level dependent (BOLD) variation. The "neurovascular coupling" between the two processes is, however, substantiated by a large number of empirical studies (Aguirre 2014). It is also important to note that dependency on secondary signals to obtain information about scientific process is not an issue unique to fMRI. As Farah points out with wilful alliteration, Chemists, Cosmologists and Climate Scientists all make inferences from secondary data. This is not, therefore, a fatal concern.

A second accusation levied at fMRI relates to the alluring simplicity of the images produced. The photograph-like appearance of brain scans presented in journal articles and, potentially, in court cases belies the enormous amount of statistical work and data processing that has been employed to reach that point. Some people get twitchy at the mere mention of statistics. As Farah notes, "*while the incorrect use of statistics in indeed misleading, there is nothing inherently misleading about using statistics… When carried out properly, statistical analyses deepen our understanding of the data and the larger reality from which they are sampled. This is the whole point of using statistical methods in any field, from cognitive neuroscience to demography.*" (Farah 2014: p. S25). The take-home message therefore is not that the use of statistics is dubious per se, but to remind us that an MRI machine being used for functional analysis of someone's brain is not actually capturing a multi-coloured photo mid-thought.

To get a feel for the scale of number-crunching involved, consider the following. As part of the fMRI scanning process, the brain is divided into sections known as volumetric pixels, or voxels for short. A typical study might involve data collection from 50,000 to 100,000 voxels in any one pass (Aguirre 2014). Re-scans then occur every 1–3 s, possibly for as long as an hour in total. That's a lot of data points!

With so many readings at such a large number of voxels, it is inevitable that random fluctuations will occur. Even if this occurs once in every thousand voxels, that would mean that something in the order of 50–100 voxels in our typical study might be randomly activated and potentially misconstrued as exhibiting a genuine response. A threshold therefore needs to be set in order to try and guard against changes that were mere chance (false positives), whilst trying to avoid missing real phenomena by setting the threshold too high (false negatives).

The danger of the former problem was brought into stark relief by an infamous experiment where researchers placed a dead Arctic Salmon in their MRI scanner. They put the salmon through the same experimental protocol that they used for their human subjects, and monitored brain activity as the fish was exposed to a series of photographs depicting human emotions (Bennett et al. 2010). They were surprised to observe apparent signals from the brain of the dead fish in response to some of the photos. Evidently these could not be detecting genuine behaviour, they must have been random changes. Carrying out simple statistical refinements enabled them to confirm these "findings" were artefacts. Nevertheless, what had started out as a joke in fact turned into an important lesson about conducting the appropriate safeguards.

As expertise in fMRI has grown, few scientists would submit a manuscript with such glaring lack of baseline control, and even fewer journals would consider publishing them. There are, however, other statistical errors that have been more persistent in the literature. Vul et al. (2009) conducted a systematic review of 53 published papers linking regions of the brain to behavioural traits. They identified that 28 fell foul of what they term the "non-independence error". In essence these studies were guilty of a degree of circularity in their reasoning. A first pass of the data was used to select interesting voxels for further study. These are likely to be areas showing different levels of activity when the research subject participates receives two contrasting stimuli. The researchers then conducted a secondary analysis of the *same* data pertaining to those selected regions in order to quantify the difference in activation. They were therefore measuring correlation in areas where they have already decided there is correlation. As Vul and colleagues point out, this does not mean to say the identified relationships are not valid (though some may be false), but the significance of the correlation is likely to have been overstated.

The apparent strength of association between certain tasks and particular regions of the brain can also be inappropriately emphasised through the choice of false-colour for modelling the activity in a final fMRI image. Vibrant colours are usually chosen to illustrate areas where a change in blood flow has been noted. Often the documented changes are relatively slight, but use of bright colour in conjunction with generous calibration of change-in-colour related to change-in-BOLD signal might make this appear more significant than is truly warranted. This has led to worries about the so-called "Christmas Tree Effect", which we will discuss in further detail in Sect. 5.2.

An aspect of some fMRI use where concerns are justified involves the process of "reverse inference". This is most prominent in the potential use of fMRI as a means of lie-detection. In reverse inference an observation is made about the activity of the brain, from which an assumption is made that the observer can know the thought processes that have led to that change (as opposed to "forward inference" in which the participant is given a known stimulus and the impact on brain response is then noted).

There is general consensus that under controlled laboratory conditions, some areas of the brain are more active when the subject is instructed to lie than when they are telling the truth (Farah et al. 2014). The correlation becomes more dubious, however, when we see activity in those sections associated with lying and assume therefore that the subject is being untruthful. It may be that other processes can lead to similar patterns of response within the brain. This is a field of research that is ripe for greater validation, after which it would still be better practice to interpret reverse inference data as offering a percentage probability of a client's truthfulness rather than a stark affirmation that they are, or are not, lying (Farah 2014).

Finally, there is a concern that apparent changes in the brain may, in fact, be influenced by other physiological processes such as breathing, heartbeat or other movement (no-one can stay perfectly still for the full duration of an fMRI scan). There is thus a theoretical danger of misinterpretation. Since, however, any well-conducted scientific study is going to look for multiple observations of the same phenomenon, it ought to be possible to disregard any one-off movement. A more significant problem might arise in so called "cognitive information testing" (Sect. 4.4.1) in which a suspect is asked questions relating to a crime scene. A guilty, but canny, defendant might be able to "game" the test by employing countermeasures such as thinking about something entirely different or making subtle movements of fingers or toes.

On balance, however, informed scholars are happy that fMRI sits comfortable alongside EEG and PET as a legitimate tool for brain study. "*Contrary to the claims of some critics, neuroimaging is not modern phrenology*" (p. S9), concludes

Aguirre (2014), adding "*It is now possible to conduct an fMRI study with great confidence in the statistical validity of the results and with a clear-eyed understanding of the assumptions on which any claims are based*" (p. S16).

Of course, confirmation of a correlation between defined regions of the brain and particular behaviours is not de facto an indication of causation, less still of compulsion. The other brain studies which have been interpreted as challenging the notion of free will are the Benjamin Libet-style investigations, in which an electrical signal, the readiness potential, appears to precede conscious awareness of a decision to move. The interpretation of such experiments has already been discussed at considerable length in Sect. 3.3.4.

It is worth reiterating here that although the readiness potential itself seems to be a real phenomenon there are diverse views regarding its significance. I am sufficiently persuaded by the critiques of Adina Roskies, Alfred Mele and others that the inevitable constraints of the experiments mean that they are not necessarily offering the insights into the timing and control of volitional actions implied by the determinists. The failure to adequately distinguish between measurement of *when* to act and *whether* to act, the potential difference between *conscious intent* and *consciousness of conscious intent*, and the inability to probe decisions made after a long period of deliberation all leave open the reality of free will, and moral responsibility.

5.1.2 Genetics

Unlike ongoing controversies about interpretation of the readiness potential, it is now beyond dispute that genetic factors play important roles in behaviour. Traditional twin and adoption studies (Sect. 3.2) which hinted at a genetic component to behaviour are being supplemented by fresh insights from molecular biology. Experiments involving animal models, corroborated by research on humans, have confirmed that there is a genetic dimension to conduct. The focus therefore has shifted from *if* genes have an effect on behaviour to identification of *how* genes are involved in behaviour and elucidation of their specific function(s).

The notion that genes *alone* are responsible for behaviour has, however, been systematically undermined by the most recent discoveries in the field. An intuitive sense that experiential stimuli and other "environmental" factors must have an influence on behaviour is now being supported at the molecular level through discoveries in the emerging fields of epigenetics and control via small non-coding RNA molecules (Sect. 3.2). As Oxford Professor Denis Noble has observed, "*Without genes we would be nothing. But it is equally true to say that with only genes we would also be nothing*" (Noble 2006: p. 44).

The need for genetic and environmental interactions (G × E) may go a long way to explaining why genome-wide association studies (GWAS) have not been as effective as anticipated at linking specific genes to particular conditions, such as diseases or behavioural patterns (Maher 2008). GWAS research involves the side-by-side comparison of the genomes of a large number of people in the hope that a spotlight will be shone on the gene or genes responsible for their shared condition. If a combination of environmental stimuli and genes are required to cause the development of a characteristic, then examination of the genes in isolation may not elicit the intended outcome. Whatever the underlying reasons for the failure of GWAS approaches to deliver the expected correlations, it is clear that a "bottom-up", causal reductionist, model in which genes blindly determine our responses, is now untenable.

We must also maintain some restraint regarding extrapolation of genetic evidence derived from animal research. In recent years, the ability to conduct comparisons of the genomes of humans and other species has facilitated the deliberate modification of interesting genes in simpler, model organisms in order to tease out information regarding the potential roles of proteins encoded by these "candidate" genes (Fry and Willmott 2011, 2012). This reductionist approach, has borne fruit, but due caution is required in interpreting findings, particularly in the area of behaviour. If a fruitfly or a mouse responds in a mechanistic manner as a consequence of a particular genetic mutation, it does not follow implicitly that humans are driven in the same way, even if they are shown to have the equivalent mutation.

As has already been noted (Sect. 3.1.2), it is the extensive development of the frontal lobe, the seat of executive functions, which marks out the human brain as different to those of other species (Miller and Cohen 2001). Crucially, a sizeable portion of this additional capacity is given over to self-inhibition, the deliberate decision to countermand inappropriate responses, and therefore not to act in certain ways (Brass and Haggard 2007; Kühn and Brass 2009). There is plenty of scope for "top-down" regulation; as Chivers (2010) notes "*there's a whole brain network associated with holding back things you shouldn't do*". Once again correlation is not causation, is not compulsion.

So, whilst a role for genes in behaviour is now established, and some brain imaging is endorsing the connections between neuroarchitecture and certain behaviours, any suggestions that moral responsibility has been killed-off have been greatly exaggerated. What bearing does this have on legal process?

5.2 Should the Science Be Used in Court and, if so, in What Ways?

As noted in Sect. 2.4, there is a widely held view that the current UK legislation regarding insanity, diminished responsibility and automatism in criminal cases is in need of major overhaul.[2] Legal experts are certainly keen to plan for a greater application of appropriate science in their proceedings. David Ormerod, heading up

[2]Although, as noted earlier, practitioners are sanguine about the likelihood of reform making it all the way to the statutes.

the Law Commission's review on fitness to plead, has argued that "*Modern criminal law should be informed by modern science and by modern psychiatric thinking*" (Ormerod 2010).

Transferring this worthy intention into legal practice is not, however, a straightforward task. Interaction of scientific evidence and legal process can lead to curious outcomes; witness the peculiar interpretations of "disease of the mind" discussed in Sect. 2.1.2. At a fundamental level, the law is ultimately looking for a binary solution—the defendant is guilty or the defendant is innocent—whereas science is more likely to present the defendant as being at a particular position on a spectrum (Martell 2009).

Both genetic and brain imaging data have now been cited in criminal cases in various jurisdictions around the world (see Sects. 4.3 and 4.4). However, their usage has been introduced in an uncoordinated manner, and serious concerns have been raise about the appropriateness of the applications to which such evidence has been put.

Although data of this kind may, in the future, become sufficiently reliable to warrant regular courtroom usage, the consensus amongst experts seems to be that the current use of neuroscientific data in determination of criminal responsibility is somewhat premature. Stephen Morse, Professor of Psychology and Law at the University of Pennsylvania, has been at the vanguard of this criticism. Somewhat tongue-in-cheek, Morse has spoken of a new pathological condition Brain Overclaim Syndrome, for which the principal symptom is a tendency "*to make claims about the implications of neuroscience for criminal responsibility that cannot be conceptually or empirically sustained*" (Morse 2006: p. 397).

Morse and other critics point to a number of substantive difficulties regarding the applicability of such evidence. By way of illustration, no pun intended, let's take the use of brain images such as those obtained using Positron Emission Tomography (PET) or functional Magnetic Resonance Imaging (fMRI).

If a court introduces neuroimage data into their proceedings it will be because they are seeking to obtain valid and relevant information regarding the role brain function might have played in the culpability of a given individual who, it is believed, has carried out a particular act at a specified time. However, interpretation of PET or fMRI scans cannot be made in isolation; there is a requirement to compare the suspect's brain against reference values. The latter are derived by averaging measurements taken from multiple different people. As a consequence there is an issue known as the "Group to individual" problem, or the G2i inference. This is awkward on both a philosophical and a pragmatic level. Philosophically, the G2i problem involves taking averaged data, whose purpose was to convey "this is generally true" and comparing it against the behaviour of one individual in one context. This has the potential to ride roughshod over all sorts of nuances in their particular case. Pragmatically, there may be defendant-specific differences in, say, brain shape or cellular neuroarchitecture that raise doubts about the legitimacy of comparison to an averaged brain.

There are other issues relating to the characteristics of the reference sample used in these kinds of comparisons. Experiments utilising these techniques have generally involved very small numbers of participants. This is, in part, due to the requirement of expensive specialist equipment. Added to this, there have been concerns about the lack of protocol standardisation and quality control in the conduct of brain scans (Ford and Aggarwal 2012).[3] Discrepancies include the methods used for calculating brain volume, and even definition of the boundaries of particular brain regions.

As a further complication to the G2i issue, it is highly unlikely that the comparator data will have been prepared especially for the court proceedings; it is more likely to the fruit of scientists' fundamental research. There is therefore a mismatch of intent; as David Faigman, Professor at the Hastings College of Law, California has noted "*While science attempts to discover the universals hiding among the particulars, trial courts attempt to discover the particulars hiding among the universals*" (Faigman 1999: p. 69).

This divergence between the purposes of science and law is all the more profound when you consider how the people who made up the comparison group were likely assembled. As I write this, there is an advert above the photocopier in my Department. Volunteers aged between 18 and 30 are being offered a small financial remuneration if they are willing to give up an hour of their time to take part in research into the effects of social gaze. This is known as a "convenience sample" and is a typical way in which subjects for psychological research are gathered. Participants are those who happen to be available, rather than a carefully selected cohort. In the context of fMRI experiments, there may be subtle differences, which we do not yet appreciate, in the brain structure or functioning between different subpopulations. It is feasible that highly-educated college students, looking to earn some cash-in-hand as volunteers in their professor's experiments, do not have 'typical' brains. Greater confidence might be generated in future trials if the control subject were more closely matched to the defendant, known in the jargon as the "ecological validity" of the sample.

Even if this was the case, however, potentially significant differences between a control group and the defendant in a trial remain apparent. A volunteer, even if they were better matched in terms of age and social experience, has little vested interest in the findings of the research. In contrast there are high-stakes issues for the defendant; the outcome might—quite literally—be a matter of life and death for her. A potential miscarriage of justice may result from the fact that real-world evidence is being compared against laboratory-based, empirical research.

Inevitably there are also issues associated with the timing of any test. A brain scan of the defendant might be conducted specifically for the trial, which may be taking place many years after the offence was committed. Is it really fair to extrapolate back from a current brain image in order to make inferences about the mental state of a suspect at the time of a crime?

---

[3]Similar concerns also mired the early uses of genetic fingerprinting, but these have since been overcome.

Some authors (e.g. Eastman and Campbell 2006; Ford and Aggarwal 2012) have expressed a worry that imaging data, especially fMRI scans, may carry unwarranted impact in a courtroom. As mentioned above, this has been termed a "Christmas tree effect" (e.g. Rushing 2014). The concern is that jurors, unfamiliar with the background science, and unaware of nuances in the conduct or interpretation of this kind of data may be dazzled by the colourful images placed before them. In consequence, the premise goes, they may be unduly swayed by apparent differences between the brain of the defendant and the "normal" brain offered as a comparator. "*An interesting neuroimage*", observes Moriarty, "*may be far more impressive to a jury than a neuroscientist's testimony about frontal lobe deficits*" (Moriarty 2008: p. 48). It was this line of reasoning which, you may recall, led the judge to ban fMRI images from the courtroom during the trial of Brian Dugan (Sect. 4.4), sanctioning instead verbal explanation of sketches.

Research into the potentially unwarranted impact of images including, but not limited to, structural and functional brain scans, has been conducted. The results to date have been rather contrary, with some studies (e.g. Gurley and Marcus 2008; Greene and Cahill 2012) suggesting that neuroimages do have particular influence on participants, whilst others (e.g. Schweitzer and Saks 2011; Schweitzer et al. 2011) found that brain images per se were not especially potent.

In a comparison of the apparent scientific credibility of written articles (unlinked to any courtroom scenario), the texts accompanied by a brain image were rated as more reliable than those featuring a graph which, in turn, were trusted more than articles without a supporting image (McCabe and Castel 2008). Earlier, Bright and Goodman-Delahunty (2006) found that showing grisly images of the victim resulted in more 'convictions' than scenarios in which these images was not included. Interestingly, however, they also found that neutral crime scene images led to a higher conviction rate than an absence of visual data, implying that *any* images connected to a crime may have influence. Although this fits into a broader story regarding the impact of visual testimony, in the current context it is worth noting two things about these experiments. Firstly, the focus of this experiment was on the *actus reus* (the conduct of the crime itself) not *mens rea* (the mental state of the accused). Secondly, no *neuro*images were included in the evidence presented.

In another study, mock jurors were asked to determine whether a defendant was guilty or not guilty by reason of insanity (NGRI) when presented with a variety of evidence (Gurley and Marcus 2008). An anatomical Magnetic Resonance Image, showing a hole in the prefrontal cortex, influenced the 'jurors' towards an NGRI verdict. Note, then, that this was a simulated trial study involving MRI data. However, no functional images of brain activity were included.

A second project using mock jurors, focussing on the 'future dangerousness' of a defendant already found guilty of a capital offence, i.e. on the penalty phase of a trail (Greene and Cahill 2012). The authors reported that participants were less likely to opt for the death penalty if they had been presented with either neuropsychological or neuroimage evidence as mitigation.

Eager to tease out these potential influences more carefully, Schweitzer et al. (2011) conducted a series of experiments looking at the impact of a variety of different images and/or expert testimony. They set up a total of six different evidence combinations, which they tested in a range of contexts (from a murder committed in conjunction with an armed robbery, through to an accidental collision between pedestrians that escalated into a brawl and a charge of assault). The evidence took the form of neuroscientific expert testimony accompanied by either a brain image, a non-brain scientific image (a graph) or a neutral image (an empty courtroom). The same empty court room image was then used alongside testimony from a clinical neuropsychologist (who emphasised physical examination of the defendant, not neurological explanations), from a clinical psychologist (who gave evidence framed in terms of mental disorder) and finally no expert witness at all.

Apparently differing from the studies described, and contrary to the a priori expectation of some of the authors, Schweitzer and colleagues did not find any specific impact of brain images over and above the influence of neuroscientific testimony. They did, however, observe that 'jurors' were more likely to look favourably on defendants if they had been exposed to explanations emphasising neuroscience rather than behavioural evidence offered by a psychologist (or no expert testimony).

This result does agree with some earlier research. For example, Weisberg et al. (2008) observed that naïve (i.e. non-expert) adults and neuroscience students found poor explanations of psychological phenomena more plausible if they were supplemented by irrelevant neuroscientific explanations (fortunately a third cohort, neuroscience experts, were able to see through the ruse). In a short review, three of the authors of Schweitzer et al. (2011) argue that the prioritisation of neuroscientific explanations is in keeping with the "lay intuition" that behavioural science represents "softer" evidence than neuroscience (Roskies et al. 2013).

A word of caution needs to be raised regarding the methodology in several of these studies. For example, McCabe and Castel (2008) sometimes exposed their participants simultaneously to more than one type of evidence, facilitating a compare-and-contrast approach which would accentuate the differences between them. Given concerns, already expressed, regarding the preponderance of college students amongst the subjects of brain scan experiments, it is ironic that the pool of mock jurors in some studies on the impact of neuroimages (e.g. Gurley and Marcus 2008; Greene and Cahill 2012) consisted entirely of undergraduate psychology majors at US universities.

Aware of the importance of studying a more representative population, Schweitzer et al. (2011) employed a professional survey service to recruit a broader cross-section of participants for an internet-based study. Of course, solving one issue can introduce another. The authors themselves were conscious of the fact that by adopting an online methodology they gained in terms of involvement of a more diverse cohort, and in larger numbers, than in other mock juror tests, but may have lost out in respect to the active engagement of the participants with the tasks. They were especially mindful of this potential criticism since their experiment recorded no significant impact of neuroimages over other neuroscientific testimony. A negative finding such as this leaves open the possibility that there was a systematic fault in the method resulting in failure to detect a real phenomenon (a so called Type II or false-negative error). Careful self-appraisal of their data, however, left them confident that the observation is valid. Later work seems to endorse this view.

The same team have subsequently employed similar methodology to investigate the effect of neuroimages in the context of both insanity defences (Schweitzer and Saks 2011) and the sentencing phase of capital trials (Saks et al. 2014). Although the former also recorded no particular influence of brain images, it seems that 'jurors' did take neuroimages into account when deciding between the death penalty and a life sentence without the possibility of parole (LWOP). This finding is therefore in agreement with Greene and Cahill (2012).

Saks et al. looked once again at a variety of testimony (clinical, genetic, and neuroscientific with and without images). They also varied whether the relevant testimony was introduced by the prosecution, as aggravation evidence in favour of the sterner punishment, or by the defence as mitigation. On this occasion, it was found that the neuroimages led to a shift towards the outcome favoured by the side introducing this evidence; i.e. more support for the death penalty if proffered by the prosecution, but endorsing LWOP if introduced by the defence.

Saks and colleagues hypothesise that the reticence of 'jurors' to be swayed into reduction of sentences in other circumstances may be down to a concern to protect the public from future risk if their decisions led to the premature re-introduction of a potentially dangerous offender into the wider community. In the sentencing phase of capital cases the public are de facto protected; the choice is between LWOP and death. As mentioned previously (Sect. 4.4), there is a broader range of acceptable evidence, and a lower threshold for inclusion, as mitigation at the penalty phase of a trail than when establishing guilt. It is also worth noting that this research may *under*estimate the potential impact of neuroscientific images in a real trial. Decisions here were made on an individual basis. With the necessity of a team of jurors to come to a unanimous (or, in some contexts, a majority) decision for the death sentence to be enacted, it would only require one or a few members of the jury to be persuaded by brain data in order for the punishment to be commuted to LWOP.

Other commentators are more sanguine about the potential issues associated with the use of neurodata. For example, Schauer (2010) has argued that scientists fretting over the limitations of fMRI data are imposing criteria from their own discipline which are unduly rigorous. Much of the legal process relies of personal testimony, recollection of individual witnesses months or years after the event, of which they may only have had a fleeting glimpse, and on the inference of connections between pieces of information which may be largely, or entirely, unwarranted. The question, Schauer argues, is whether the evidence from brain imaging is at least as reliable as some of these other methods. He notes that "*the evaluative standard to be used by the law, even when it is science that is being evaluated, must be based on law's goals, law's purposes, and law's structures*" (Schauer 2010: p. 1219).

Additional controversy regarding the use of scientific data in the courtroom has focused on the role played by "expert" witnesses. As noted previously (Sect. 4.2), US Federal Rules of Evidence number 702 requires an expert witness to be "*qualified as an expert by knowledge, skill, experience, training or education*" (quoted by Moriarty 2008: p. 36). The witness must be speaking of a discipline in which they have specialist knowledge and be discussing evidence that is reliable both in the methodology used but also, significantly, in the *interpretation* being placed upon that evidence. There is particular concern that some psychiatrists making pronouncements on the validity of brain imaging or genetic evidence do not have the appropriate background in the conduct of the techniques that gave rise to the data.

Finally, the confirmation of an underlying genetic or physiological explanation for antisocial behaviour does not lead inevitably to exemption from responsibility. "*Even if science does develop to show correlations of particular genes, or types of brain state, with aggression, will that necessarily infer diminished or absent moral or criminal responsibility?*" (Eastman and Campbell 2006: p. 311). Morse has been particular animated about the leap from explanation to exoneration, which he has identified as the 'fundamental psycholegal error'. "*Believing that causation per se mitigates or excuses responsibility*" he argues "*is the most pernicious confusion bedeviling the attempt to relate scientific findings to criminal responsibility*" (Morse 2011: p. 379).

5.2.1 The Legal Applicability of Biological Evidence

So, *should* biological evidence be admissible in court? We may agree with the sentiment of Sinnott-Armstrong and Nadel (2011: p. xi) that "*Courts do not need to settle the issue of determinism before they put criminals in jail*". If, however, brain scans reveal significant abnormality in the prefrontal cortex of a suspect, damage which is likely to have had a detrimental effect on their capacity for self-control, can we in good conscience disregard this? Or suppose that an individual is shown to have a genotype with well-corroborated links to some kind of psychopathic behaviour. Are we going to simply ignore this information?

Any advocate for justice ought to be in favour of the introduction of legitimate and relevant evidence at the appropriate point in legal proceedings. The devil, of course, is in the detail and two caveats were present within that statement: *Is* the evidence legitimate and relevant? And *when is* the appropriate point for such evidence to be proffered? These questions are linked, but I believe the clearest route forward comes by tackling them in reverse order.

As we have seen (Sect. 4.3.1), there is precedent for the use of genetic data during the liability phase of a trial (e.g. for Bradley Waldroup) and in the sentencing phase in the trail (e.g. for Adbelmalek Bayout). On present evidence, I concur with Baum (2013) that it is within the sentencing phase, i.e. *after* determination of guilt, that any such evidence might have a role to play.

If a crime has been committed it is important for the functioning of society that the agent is dealt with appropriately. Even if neurobiological evidence moves on from establishing correlation to confirming causation this is unlikely to be so compelling as to fully exonerate the perpetrator. As a recent report by the

Presidential Commission for the Study of Bioethical Issues concluded "*[E]ven if we can agree that moral decisions result from neurological chain reactions, it does not follow that the individuals who make those decisions should not be held ethically or legally responsible for them.*" (Bioethics Commission 2015: p. 105).

If biological circumstances (or indeed other factors) are, at some future time, shown indisputably to have had significant influence on the wrongdoer this can be reflected during sentencing. Of course this takes us into the territory of another major philosophical dilemmas in criminal justice: the purpose of punishment. Is a sanction applied as retribution against the offender, opportunity for their rehabilitation, as a warning to deter other would-be criminals, as a source of comfort to the victim, or some combination of the above? Is the prime motivation for setting a custodial, or indeed a capital, sentence payback on the offender, or protection of the public? These questions are worthy of another book in their own right [indeed readers interested in pursuing that topic in greater detail are recommended Hudson (2003)].

For our current purposes it ought to be sufficient to assume that the core priority of punishment is to reduce the future occurrence of crime. It is just worth reiterating, however, that biological evidence of causation, if confirmed, is really a double-edged sword. Rather than being grounds for a *reduced* sentence, might an apparent defence "I couldn't help it, I've got bad genes" in fact be evidence that the individual has a greater likelihood of re-offending, thereby necessitating a *longer* custodial sentence?

The second question, then, is whether the evidence being proffered is genuinely the fruit of "mature and verified science" (Bioethics Commission 2015: p. 102) utilised in an appropriate way. Despite the fact that some neurobiological evidence has been cited during criminal cases, I would argue that, at this present time, few of these uses have been sufficiently robust to give confidence regarding expansion of this provision in the near future.

As we have seen above, the fundamental science is essentially sound, it is the application of that science to criminal justice which seems somewhat premature. The exception to this statement would appear to be the impact of research into the adolescent brain. Commentators generally seem in agreement that evidence for the delayed development of the prefrontal cortex, involved in rational decision-making and self-control, means adolescents are not simply "mini-adults" (Galván 2014; Blakemore 2012). This neuronal plasticity contributes to the impulsivity of youth, their capacity to be led astray, but also to the greater potential for reform than might be the case for older offenders. Both the reduced culpability and increased scope for rehabilitation[4] therefore warrant the recent changes in the treatment of juveniles in the American courts. The criticism relating to adolescent brains might be that other jurisdictions have so far failed to action the necessary response to this research.

In other spheres, the concern remains that too much attention has been given to science whose legal application has yet to be substantiated. With regard to brain

[4]Of course the ways to achieve rehabilitation of young offenders is another matter entirely.

imaging, different techniques have different hurdles to overcome. However, the need to improve the ecological validity of control populations is common to all methods. The current practice of comparing a defendant against reference data derived from carefully-orchestrated laboratory conditions, and (likely) involving well-educated students as the research subjects, needs to be replaced by analysis of more appropriate, "real life" populations.

Regarding specific approaches, Ben-Shakhar and Kremnitzer (2011) suggest that EEG for critical information testing passes three out of the four Daubert standards (testability, exposure to peer review and general acceptance by the relevant expert community) but questions remain about the final qualification, known error rate. In particular this methodology may be susceptible to frequent false-negative rates if, for example, a habitual burglar does not take sufficient notice of one property amongst the many he has raided during his career. Incidentally, some courts have also been reticent to allow EEG-based evidence on more philosophical grounds, seeing it as usurping the role of jurors in determining the credibility of defendants, and other witnesses (Rosenfeld et al. 2013). An expectation that, for justice sake, the best available evidence ought to be made available must surely trump this more semantic view.

For fMRI, the reverse inference issue remains a serious concern; substantial work needs to be done on validation of connections between specific regions of the brain and particular thought processes. Progress has been made in respect of non-independent analysis of data, but this too is an area in which researchers must not be complacent.

With regard to the use of genetics, allusion to hypothetical genes (as in the case of Gary Cossey, Sect. 4.3.1) is of no benefit to anyone. Rather than standing as an example of the potential for the use of genetic evidence in criminal law, I believe the case of Abdelmalek Bayout (Sect. 4.3.1) should also serve as a warning regarding potential abuses. Too many details regarding environmental influences on Bayout did not fit with the prevailing scientific understanding of the significant genotype.

A straightforward correlation between a small set of genes and behaviour has yet to emerge; it is possible that it may never do so. However, despite the limited contribution of GWAS approaches to date, it may yet be that genomics will provide a significant connection. Ever plummeting cost of genomics makes it feasible to add a molecular dimension to traditional family tree (pedigree) studies of patterns of behaviour. In this way the influences of genes, of environmental stimuli and a combination of the two might be teased out more precisely.

5.2.2 Achieving a Just Outcome in Trials Employing Neurobiological Evidence

The aim of the legal process is to reach the correct verdict; to achieve a fair outcome for both victim and perpetrator. Doing so involves both the promotion of justice and the reduction of injustice. How should corroborated scientific evidence be treated?

It may be asked, with some justification, whether it is appropriate to take into account genetic influences when other factors, biological or environmental may have played as large or an even greater role in determining somebody's action. Consider, for example, the part allegedly played by *MAOA* genotype.

By turn of serendipitous events, *MAOA* has come to be the best studied behavioural gene (Sect. 3.2.3). However, if the Dutch kindred had not had a mutation with such a profound effect, or if they had not been the subjects of Brunner et al.'s (1993) analysis, then this particular gene might not have been in the spotlight. If Cases et al. (1995) had not fortuitously knocked out the equivalent gene whilst conducting random mutagenesis in mice, then the story might not have developed. And if Caspi et al. (2002) had not already spent over 25 years conducting a long-term study of abuse as a risk-factor to subsequent aggression then the importance of genetic and environmental influence acting together might not have been so readily established.

The key point here is this; just because *MAOA* is the most studied gene, it does not necessarily follow that it is the most important genetic influence upon behaviour. Can we legitimately take into account an individual's *MAOA* genotype in reducing their sentence when another person with an as-yet-unidentified mutation in a different gene may actually be under greater genetic influence to be aggressive?

Against this, it might be argued that you can only take into account the evidence that you *do* have. By analogy, you cannot include the testimony of an eyewitness who would have confirmed a suspect's alibi if the witness is not located in time to give evidence at the trial. This does not mean, however, that eyewitness testimony cannot be used in a different trial where it would be applicable. Similarly, it would not be fair to exclude MAOA data (once appropriately validated) even though there may be an unknown genetic or physiological influence having as great or greater bearing on a different defendant.

Of course, poor conduct of any assessment giving rise to biological evidence can bring into disrepute the method employed and undermine wider confidence its usage. For example, the fMRI lie-detector testing of Lorne Semrau (Sect. 4.4.1), which was carried out a third time after the first two readings disagreed, is not best practice. "Two out of three" is not an adequate strike rate to instil confidence in the legitimacy of the findings. It is important that the highest possible standards are employed in the gathering of any data.

Some potential sources of injustice are not easily reconciled. Due to the specialist machinery involved, PET and fMRI are going to remain expensive methods of garnering evidence, and all the more so if, as recommended above, bespoke control group data is going to need to be generated. There are therefore issues of uneven access based on the financial limitations of some defendants.

A key factor in improving the appropriate usage of biological evidence is going to be the development of a more scientifically literate judiciary and jurors. Two significant reports examining the implications of emerging knowledge in neuroscience upon the legal system have recently been produced. In 2011, the UK Royal Society commissioned a report *Brain waves module 4: neuroscience and the law* (Royal Society 2011) and the USA the Presidential commission for the study of bioethical issues (Bioethics Commission 2015) looked more generally at the interface of neuroscience, ethics and society, including the law.

A theme shared in the recommendations emerging from both reports was the need for appropriate training of legal professionals and others in the basics of genetics and brain science. The Royal Society advocated the incorporation of modules on pertinent science topics as part of the core training of law students at university and the establishment of formal Continuous Professional Development requirements for lawyers, judges and probation officers. They also recommended that an international conference involving scientists and key legal practitioners should take place every three years in order to disseminate the latest findings.

In some senses the USA has already moved further in this regard. With the support of over 15 million dollars from the John D. and Catherine T. MacArthur Foundation, a Research Network on Law and Neuroscience has been established. The Network (www.lawneuro.org) offers a burgeoning collection of pertinent resources for all interested parties, as well as organising law and neuroscience courses at universities in America and overseas. Elsewhere Owen Jones, Director of the Network, has also stressed the reciprocal importance of training scientific expert witnesses so that they have appropriate knowledge of the different expectations of evidence in a legal context as opposed to empirical research (Jones et al. 2013).

Greater acceptance that diverse factors have played a role in influencing someone's behaviour may necessitate the development of more nuanced penalties. For example, might there be a case for saying that criminality associated with a brain tumour should be worthy of more sympathetic consideration that a genetic influence (Peter Clarke, personal communication)? The justification for such a decision could be framed as follows. Genetic effects are internal and likely to be long-term. maybe even life-long. As such, there may have been opportunities to develop compensatory control mechanisms. A tumour, on the other hand, can develop suddenly such that there has not been time to learn to limit its influence. By legal precedent, the *Charlson* case, it is also an "external" rather than an internal factor. There is also far greater likelihood or reversing the antisocial behaviour, via removal of the tumour, than would be possible to counteract the influence of genetics.

5.3 Conclusion

I believe it is inevitable that neuroscientific and behavioural genetic evidence will play a more significant part in criminal law. However, it is crucial that the timing of this expanded usage is apposite. The complexities underlying the interaction of genetics, of neurobiology and of environmental influences are not adequately understood at present to endorse their routine used for sentencing, let alone determination of guilt. In due course there may be justification for accepting that some individuals have had to struggle against harder biological heritage than others and, in consequence, might be permitted an appropriate reduction in their punishment.

In the meantime, even enthusiasts for greater emphasis on neurobiological factors need to add their own sandbags against the rising tide, to delay the arrival of this wave until a time when it can be appropriately channelled. Up until that point, there remains insufficient evidence to warrant the suggestion that any biological factors are of such significant influence that moral responsibility has been abrogated.

References

Aguirre G.K. 2014. Functional neuroimaging: technical, logical and social perspectives. *Hastings Center Report*, March–April 2014, S8–S18.

Ariely, D., and G.S. Berns. 2010. Neuromarketing: the hope and hype of neuroimaging in business. *Nature Reviews Neuroscience* 11: 284–292.

Baum, M.L. 2013. The monoamine oxidase A (MAOA) genetic predisposition to impulsive violence: is it relevant to criminal cases? *Neuroethics* 6: 287–306.

Bennett C.M., A.A. Baird, M.B. Miller, and G.L. Wolford 2010. Neural correlates of interspecies perspective taking in the post-mortem arctic salmon: an argument for proper multiple comparisons correction. *Journal of Serendipitous and Unexpected Results* 1:1–5. Available at http://pages.vassar.edu/abigailbaird/files/2014/06/bennett_salmon.pdf. Last accessed 18th Aug 2015.

Ben-Shakhar G., and M. Kremnitzer 2011. The concealed information test in the courtroom: legal aspects. In: *Memory detection: theory and application of the concealed information test*, eds. Verschuere, B., Ben-Shakhar, G., Meijer, E., 276–292. New York, USA: Cambridge University Press.

Bioethics Commission. 2015. Gray matters (volume 2): topics at the intersection of neuroscience, ethics, and society. Available online at http://bioethics.gov/sites/default/files/GrayMatter_V2_508.pdf. Last accessed 8th Aug 2015. Washington DC, USA: Presidential Commission for the Study of Bioethical Issues.

Blakemore, S.-J. 2012. Imaging brain development: the adolescent brain. *Neuroimage* 61: 397–406.

Brass, M., and P. Haggard. 2007. To do or not to do: the neural signature of self-control. *The Journal of Neuroscience* 27: 9141–9145.

Bright, D.A., and J. Goodman-Delahunty. 2006. Gruesome evidence and emotion: anger, blame, and jury decision-making. *Law and Human Behavior* 30: 183–202.

Brunner, H.G., M. Nelen, X.O. Breakefield, H.H. Ropers, and B.A. van Oost. 1993. Abnormal behavior associated with a point mutation in the structural gene for monoamine oxidase A. *Science* 262: 578–580.

Cases, O., I. Seif, J. Grimsby, P. Gaspar, K. Chen, S. Pournin, U. Muller, M. Aguet, C. Babinet, J. C. Shih, and E. De Maeyer. 1995. Aggressive behavior and altered amounts of brain serotonin and norepinephrine in mice lacking MAOA. *Science* 268: 1763–1766.

Caspi, A., J. McClay, T.E. Moffitt, J. Mill, J. Martin, I.W. Craig, A. Taylor, and R. Poulton. 2002. Role of genotype in the cycle of violence in maltreated children. *Science* 297: 851–854.

Chivers T. 2010. Neuroscience, free will and determinism: 'I'm just a machine'. *Telegraph*, 12th Oct 2010. Available online at http://www.telegraph.co.uk/science/8058541/Neuroscience-free-will-and-determinism-Im-just-a-machine.html. Last accessed 8th August 2015.

Eastman, N., and C. Campbell. 2006. Neuroscience and legal determination of criminal responsibility. *Nature Reviews Neuroscience* 7: 311–318.

Economist. 2006. Free to choose? *The Economist*, 19th December 2006. Available online at http://www.economist.com/node/8453850. Last accessed 1st Sept 2012.

Faigman, D.L. 1999. *Legal alchemy: the use and misuse of science in the law*. New York: St Martin's Press.

Farah, M.J. 2014. Brain images, babies, and bathwater: critiquing critiques of functional neuroimaging. *Hastings Center Report*, March–April 2014, S19–S30.

Farah, M.J., B. Hutchinson, E.A. Phelps, and A.D. Wagner. 2014. functional MRI-based lie detection: scientific and societal changes. *Nature Reviews Neuroscience* 15: 123–131.

Ford, E., and N. Aggarwal. 2012. Neuroethics of functional neuroimaging in the courtroom. In *Neuroimaging in forensic psychiatry: from the clinic to the courtroom*, ed. J.R. Simpson, 325–340. Chichester, UK: Wiley.

Fry, A., and C. Willmott. 2011. "Model organisms in biomedical research" (video). Available online at http://youtu.be/Jj5QlYIE66w. Last accessed 18th Oct 2012.

Fry, A., and C. Willmott. 2012. The power of comparative genomics (video). Available online at http://youtu.be/mU9ROpm6d70. Last accessed 18th Oct 2012.

Galván, A. 2014. Insights about adolescent behavior, plasticity, and policy from neuroscience research. *Neuron* 83: 262–265.

Glimcher, P.W., and A. Rustichini. 2004. Neuroeconomics: the consilience of brain and decision. *Science* 306: 447–452.

Greene, E., and B.S. Cahill. 2012. Effects of neuroimaging evidence on mock juror decision making. *Behavioral Sciences and the Law* 30: 280–296.

Gurley, J.R., and D.K. Marcus. 2008. The effects of neuroimaging and brain injury on insanity defenses. *Behavioral Sciences and the Law* 26: 85–97.

Gutting, G. 2011. What makes free will free? *New York Times*, 19th Oct 2011. Available online at http://opinionator.blogs.nytimes.com/2011/10/19/what-makes-free-will-free. Last accessed 2nd Sept 2012.

Henry, S., and D. Plemmons. 2012. Neuroscience, neuropolitics and neuroethics: the complex case of crime, deception and fMRI. *Science and Engineering Ethics* 18: 573–591.

Hudson, B.A. 2003. *Understanding justice: an introduction to ideas, perspectives and controversies in modern penal theory*, 2nd ed. Buckingham: Open University Press.

Jones, O.D., A.D. Wagner, D.L. Faigman, and M.E. Raichie. 2013. Neuroscientists in court. *Nature Reviews Neuroscience* 14: 730–736.

Kühn, S., and M. Brass. 2009. When doing nothing is an option: the neural correlates of deciding whether to act or not. *Neuroimage* 46: 1187–1193.

Lopes da Silva, F. 2013. EEG and MEG: relevance to neuroscience. *Neuron* 80: 1112–1128.

Maher, B. 2008. Personal genomes: the case of the missing heritability. *Nature* 456: 18–21.

Marks, J.H. 2010. A neuroskeptic's guide to neuroethics and national security. *American Journal of Bioethics: Neuroscience* 1: 4–12.

Martell, D.A. 2009. Neuroscience and the law: philosophical differences and practical constraints. *Behavioral Sciences and the Law* 27: 123–136.

McCabe, D.P., and A.D. Castel. 2008. Seeing is believing: the effect of brain images on judgments of scientific reasoning. *Cognition* 107: 343–352.

Michel, C.M., and M.M. Murray. 2012. Towards the utilization of EEG as a brain imaging tool. *Neuroimage* 61: 371–385.

Miller, E.K., and J.D. Cohen. 2001. An integrative theory of prefrontal cortex function. *Annual Review of Neuroscience* 24: 167–202.

Moriarty, J.C. 2008. Flickering admissibility: neuroimaging evidence in U.S. courts. *Behavioural Sciences and the Law* 26: 29–49.

Morse, S.J. 2006. Brain Overclaim Syndrome and criminal responsibility: a diagnostic note. *Ohio State Journal of Criminal Law* 3: 397–412.

Morse, S.J. 2011. Genetics and criminal responsibility. *Trends in Cognitive Sciences* 15: 378–380.

Noble, D. 2006. *The music of life*. Oxford: Oxford University Press.

Ormerod, D., 2010. Quoted in law commission press release. *New rules to decide who is fit to stand trial*. 27th Oct 2010. No longer available on the Law Commission website, but archived at http://www.epolitix.com/stakeholder-websites/press-releases/press-release-details/

newsarticle/new-rules-to-decide-who-is-fit-to-stand-trial///sites/law-commission/. Last accessed 6th Sept 2012.

Poole S. 2012. Your brain on pseudoscience: the rise of popular neurobollocks. *New Statesman*, 6th Sept 2012. Available online at: http://www.newstatesman.com/culture/books/2012/09/your-brain-pseudoscience-rise-popular-neurobollocks. Last accessed 18th Aug 2015.

Rachul, C., and A. Zarzeczny. 2012. The rise of neuroskepticism. *International Journal of Law and Psychiatry* 35: 77–81.

Rosenfeld, J.P., X. Hu, E. Labkovsky, J. Meixner, and M.R. Winograd. 2013. Review of recent studies and issues regarding the P300-based complex trial protocol for detection of concealed information. *International Journal of Psychophysiology* 90: 118–134.

Roskies, A. 2006. Neuroscientific challenges to free will and responsibility. *Trends in Cognitive Sciences* 10: 419–423.

Roskies, A.L., N.J. Schweitzer, and M.J. Saks. 2013. Neuroimages in court: less biasing than feared. *Trends in Cognitive Sciences* 17: 99–101.

Royal Society. 2011. *Brain waves module 4: neuroscience and the law*. London: The Royal Society.

Rushing, S.E. 2014. The admissibility of brain scans in criminal trials. *Court Review* 50: 62–69.

Saks, M.J., N.J. Schweitzer, E. Aharoni, and K.A. Kiehl. 2014. The impact of neuroimages in the sentencing phase of capital trails. *Journal of Empirical Legal Studies* 11: 105–131.

Schauer, F. 2010. Can bad science be good evidence? *Neuroscience, Lie Detection and Beyond, Cornell Law Review* 95: 1191–1219.

Schweitzer, N.J., and M.J. Saks. 2011. Neuroimage evidence and the insanity defense. *Behavioral Sciences and the Law* 29: 592–607.

Schweitzer, N.J., M.J. Saks, E.R. Murphy, A.L. Roskies, and W. Sinnott-Armstrong. 2011. Neuroimages as evidence in a mens Rea defence: no impact. *Psychology, Public Policy, and Law* 17: 357–393.

Sinnott-Armstrong, W., and Nadel, L. 2011. Introduction. In *Conscious will and responsibility*, ed. W. Sinnott-Armstrong and L. Nadel, xi–xvi. New York, USA: Oxford University Press.

Stafford T. 2015. Why do we intuitively believe we have free will? BBC website, 7th Aug 2015. Available online at http://www.bbc.com/future/story/20150806-why-your-intuitions-about-the-brain-are-wrong. Last accessed 8th Aug 2015.

Vul, E., C. Harris, P. Winkielman, and H. Pasher. 2009. Puzzlingly high correlations in fMRI studies of emotion, personality, and social cognition. *Perspectives on Psychological Science* 4: 274–290.

Weisberg, D.S., F.C. Keil, J. Goodstein, E. Rawson, and J.R. Gray. 2008. The seductive allure of neuroscience explanations. *Journal of Cognitive Neuroscience* 20: 470–477.

Zimmer, L., and A. Luxen. 2012. PET radiotracers for molecular imaging in the brain: past, present and future. *Neuroimage* 61: 363–370.

The manufacturer's authorised representative in the EU is Springer Nature Customer Service Centre GmbH, Europaplatz 3, 69115 Heidelberg, Germany. If you have any concerns regarding our products, please contact ProductSafety@springernature.com

Printed and bound by CPI Group (UK) Ltd, Croydon, CR0 4YY
25/09/2026
02222857-0015